THE DIFFERENCE

Bringing an Organization
Back to Life

TIM PASSMORE

Published by Outcome Publishing, 818 West Diversey Parkway, Suite W, Chicago, Illinois 60614
www.outcomepublishing.com

ISBN: 978-0-9826127-0-5

CONTENTS

PREFACE

*T*he lessons I share in this book have become very real to me over the past several years. They are not a collection of ideas that might work; they are truths that do work. They have been proven effective in producing change in organizations, whether "for profit" or "not-for-profit."

I have had the privilege of working in both "for profit" and "not-for-profit" organizations. After graduating from college, my aspirations were to work in a large corporation and begin my professional journey. This was realized when I was hired by one of the largest utility corporations in America as a marketing representative. My responsibility was to assist in the recruitment of manufacturing corporations in an effort to encourage them to build facilities within our service territory. This was an amazing experience as I learned much about corporate life.

My professional journey took a major turn when I decided three years later to enter into another field, work within the church. This was a life I was accustomed to, having grown up in the home of a pastor who was a very successful leader.

I have discovered through my involvement in both "for profit" and "not-for-profit" organizations that the same principles apply to both related to leadership. In studying the writings of influential figures in leadership education and through my personal experience it has become apparent to me that these principles are critical for success. This especially became obvious to me after taking on a specific leadership position.

When I arrived as pastor at Woodland, a church in Bradenton, Florida, I realized that there were many challenges to address. At the time of my arrival, Bradenton was one of the fastest growing areas in the United States. Many people were moving into the community and were looking to connect to religious organizations to further their spiritual interest. We were committed as a group of people to reach them and to begin meeting their needs.

Our organization needed a plan and an environment that encouraged cooperation among our members as we reached out into the community for us to accomplish our goal. Everyone needed an understanding of what we should do and also needed the proper passion to accomplish the plan. We would have no significant results unless we addressed areas of concern.

These were great pressures for a young leader and they began to overwhelm me. In spite of this, the church began to experience growth as we were reaching more and more people. As we grew, it became apparent that my leadership abilities would need to improve in order to maintain a healthy organization. Because of this, I made the decision to pursue a doctorate degree in leadership and administration to learn how to be more effective. Gaining this education proved to be a critical time as I was stretched as an individual and as someone who influenced a large group of people. This new knowledge assisted me in understanding and uncovering leadership principles that make "THE DIFFERENCE." They would be the foundation for my influencing others to accomplish something we considered to be of great significance.

My successes and failures along with my studies related to organizational health are the backdrop for writing this book. This is not a book meant only for leaders within religious or non-profit organizations, it is meant for leaders within any organization. The same principles apply to all.

Some of the best leadership advice I've received is simply to do four things: seek good council, make wise choices, refrain from complaining, and be a continual learner. You have the opportunity to become a continual learner by reading the following pages. My hope is that it will focus your attention on things that you can change in a positive way, encouraging you to concentrate on healthy action and away from unhealthy attitudes. Having the right perspective about our involvement in organizational life influences us to seek out good advice from those who have knowledge that can assist us, leading us to make wise choices. I encourage you to take the needed time to consider the information in this book as you grow in your leadership. You can make a difference! It's time to learn how.

INTRODUCTION

*W*ill it hit us? Many residents in our area were asking this question as the weather experts continually forecasted the landfall of hurricane Charley. My children had never experienced a hurricane before and were actually excited about the event. This may sound morbid, especially when you understand the destructive nature of a category-four storm. They weren't excited about the hurricane itself; the hurricane merely allowed them to have a day off from school.

I didn't share the enthusiasm of my children. I remembered growing up on the Gulf Coast of Florida and experiencing hurricanes during my childhood. I saw the evidence of their destructive power. Who can forget the images shown on our television screens after hurricane Katrina? I still remember the pictures of homes that were reduced to piles of rubble by the storms.

Great advances in technology help forecast hurricanes today. My wife and I watched the telecast hosted by local newscasters as they began to notice on radar that the storm was turning. It moved toward the east, directly into Charlotte Harbor, approximately 50 miles south of our home. Our community was spared, but unfortunately the city of Punta Gorda was not. Many houses in that area looked as if someone had plowed through them on a bulldozer, with no regard for the possessions within their walls.

Thousands of homes were declared uninhabitable by the Federal Emergency Management Agency, while others survived the winds and rain with little damage. Why did they survive? Those who have lived in Florida for some time know that the hurricane building codes have changed over the years. More recently, the construction of homes has followed a building plan that produces a structure that can withstand the force of a violent storm. After all, "crashed" homes are not pretty sites. Seeing them motivates those who live in high risk areas to prepare for the storms that may one day come.

How does this relate to organizational life? Unfortunately, there are many that have crumbled. They have been reduced to rubble. This can be quite tragic, especially understanding the makeup of organizations. They are made of people and we have the opportunity to impact them in a positive way.

In her book *John Paul the Great*, Peggy Noonan described an encounter she had with an American CEO. She wrote,

> I am talking with the head of a mighty American corporation. We're in his window-lined office, high in midtown Manhattan. The view – silver sky-scrapers stacked one against another, dense fine-lined, sparkling in the sun – is so perfect, so theatrical, it's like a scrim, like a fake backdrop for a 1930's movie about people in tuxes and tails... The CEO tells me it is "annual report time" and he is looking forward to reading the reports of his competitors. Why? I asked. I wondered what he looks for specifically when he reads the reports of the competition. He said he always flipped to the back to see what the other CEOs got as part of their deal – corporate jets, private helicopters, whatever.

"We all do that," he said. "We all want to see who has what." He was a talented and exceptional man, and I thought afterward that he might, in an odd way, be telling me this about himself so I wouldn't be unduly impressed by him. But what I thought was, *It must be hard for him to keep some simple things in mind each day as he works.* Such as this: A job creates a livelihood, a livelihood creates a family, a family creates a civilization. Ultimately, he was in the civilization-producing business. Did he know it? Did it give him joy? Did he understand that that was probably why he was there? I thought: *This man creates the jobs that create the world in which we live. And yet he can't help it, his mind is on the jet.*[1]

We really are in the civilization building business. The way we treat people determines how well we accomplish this goal.

People are important for another reason. They are the key to bringing about maximum results. This is important to remember because the influence of those within the organization can lead to either success and strength or failure and weakness. Our desire should be to produce organizational partners who use their abilities to have the greatest effect possible. Organizations need people who won't buckle under pressure; they need those who will ensure that the purpose is fulfilled. We also need those who are willing to make the sacrifices that are necessary to achieve the cause, whatever it may be.

The health of organizations today is dependent upon leaders and partners who have "rock solid" strength and are dedicated to live out their passion for accomplishing their goals. To ensure that this happens requires our dealing well with pressure and moving forward in a way that benefits organizational life. We must work together to prevent it from crashing down. Many organizations have experienced success in the past but now find themselves in a strained position, losing their momentum and effectiveness. They seem to be moving backwards instead of forwards. What caused this to happen? What can be done to overcome this? What makes "THE DIFFERENCE?"

STAGES OF ORGANIZATIONAL HEALTH

Jim Collins, in his book *How the Mighty Fall*, wrote about why companies fail. He teaches that companies move through stages which ultimately lead them to become irrelevant or to die. The stages include:

Stage 1: Hubris Born Of Success. During this stage, there is much momentum. Those who make up the organization begin to believe that they are invincible and can do no wrong. They begin to lose sight of the factors that caused them to initially experience success.

Stage 2: Undisciplined Pursuit Of More. During this stage, the company begins to believe that more is better. They work to achieve more growth and more recognition. Problems occur when they grow beyond their ability to secure key personnel for needed positions. They set themselves up for a fall.

Stage 3: Denial Of Risk And Peril. During this stage, there are internal warning signs but leaders ignore them. They discount the negative and amplify the positive. Leaders begin to blame external factors for their problems rather than taking personal responsibility.

Stage 4: Grasping For Salvation. During this stage, the company experiences sharp decline which is noticeable. They look for a quick solution. Collins calls these "saviors." He shares that they may include "a charismatic visionary leader, a bold but untested strategy, a radical transformation, a dramatic cultural revolution, a hoped-for blockbuster product, a 'game changing' acquisition, or other silver bullet solutions." The use of these may produce positive results initially, but they are not sustained.

Stage 5: Capitulation To Irrelevance Or Death. During this stage, the attempts to turn the company around ultimately fail and it begins to spiral downward. The organization begins to lose strength and hope is gone.[2]

The stages mentioned above are very personal to me and my leadership. Woodland has changed dramatically since I became leader 14 years ago. At first, it was almost as if we could do no wrong. Our organization began to grow quickly and momentum was building. We began to expand ministries and add new ways in which we could reach our community. Although the motivation for this was correct, it caused many challenges. Our expansion of ministries outpaced our ability to fill the needed positions of leadership necessary to oversee those ministries. The result – frustration!

We found ourselves in a new dynamic. There was slower growth, a lack of momentum, and more pressure put on volunteers. We began to look at why these issues were taking place. Attention was put on a changing community; one that was no longer growing because of difficult economic times. We spoke of other churches that had begun in our area; churches that had been effective in reaching people within the community. Both were legitimate issues that had affected who we were, however, they didn't tell the whole story. One critical factor remained. There were underlying leadership failures which occurred because of our lack of attention placed on fundamental areas that affect organizational health.

The story does not stop there. Because of stagnancy in the church, new ways in which to do ministry were implemented in hopes that they would be quick fixes. For example, dealing with a lack of volunteers led to the re-organization of some ministries which caused further challenges. These changes left members feeling disconnected from the church.

I don't know if you've noticed this as I have shared our story, but we found ourselves in *Stage 4: Grasping For Salvation*. This is not a comfortable position to be in. So what was the answer? It was to get back to the principles of leadership that make "THE DIFFERENCE."

As I write this book, we are still in the process of moving toward organizational success once again. It's not easy and it takes time. The good news is that it can bring positive change. Already, there is a greater sense of motivation as we seek to improve who we are as leaders and as a team determined to fulfill our mission.

REGAINING FOCUS

Is your organization in one of the stages? Why does all of this occur? As we have discovered, it goes back to Stage One when leadership loses sight of the disciplines that brought about their success to begin with. A critical factor in regaining success is to reestablish these disciplines once again. This book is about the disciplines that must be present for the organization to stand and avoid a crash.

Leaders need to think of themselves as builders who practice disciplines that bring organizational strength. They are to be people who...

- Never lose sight of what makes or breaks an organization.
- Construct an organization that stands when challenges come.
- Adapt when changes occur.
- Are aware of the danger of moving into the stages that lead to the organization's failure.

Success is also dependent upon our knowing our roles and how we complement one another. This brings unity and a sense of partnership and keeps our focus on what positively influences the organization. I refer to organization workers or members as "partners." We are in this together as we move forward in ensuring that our vision becomes reality. This requires our appreciating one another and understanding the significant impact others can have because of their abilities. The goal is to work together to achieve an outcome that matters.

FOUNDATIONAL POSTS

Leaders and partners are to construct organizations that are effective. For this to occur, they must keep in mind the principle of "cause and effect." For any effect, there is a cause. In other words, there is a reason for the OUTCOME. There are several elements that cause organizational success and bring strength to our structure. These elements become the basis for an organizational building plan and form the framework for the construction process. Each gives us stability and must be present for us to succeed. If an element of the framework is missing or is weak, it changes the OUTCOME of what we are able to achieve.

Let's illustrate it this way. Foundational posts hold up a beach home along the ocean. If one post is destroyed from the pressure of a wave, the home is weakened. The question is, "Why did the post give way?" The answer - Poor construction! It wasn't able to withstand the pressure.

A few years ago, Hurricane Ivan came barreling through the Pensacola area causing massive destruction. This was personal to me having grown up there and also living in the community during the first part of my professional career. Just a few weeks after the storm, my wife and I visited the destruction zone with some close friends who gave us a tour of some of the damage. We saw an amazing sight. In one particular neighborhood, two homes had been built next to one other on the Inter-coastal Waterway. Both looked very similar on the outside. We know this because we lived nearby after graduating from college. These homes were along our regular walking and biking path. Now, things were much different. In the aftermath of the storm, one house remained standing with little damage and the other was gone. I mean literally gone! There was only a slab of concrete which marked where it once stood. It had been pushed down by the storm and washed away by the power of the waves.

Ivan revealed the truth about the structures. Although they looked similar, their foundations were quite different. One was built on round posts which allowed the pressure to move easily around the foundation and the other was built with posts made from square cinder blocks. The water had no way to navigate around the flat surface. The pressure of the waves, brought in by a rising tide, pushed them down and the entire house was compromised. Complete destruction was the result all because of one flaw - poor "post" construction.

Let's think about organizations. There are certain factors that keep an organization standing. We'll also refer to them as "foundational posts." Their critical role in organizational health requires us to make certain that each is constructed well. If a post is poorly constructed, it won't be able to withstand the pressure, causing the organization to be compromised. The OUTCOME it produces is affected in a negative way. The area of vulnerability, if not addressed and strengthened, has a devastating affect – destruction! Because this is true, it is important to identify what they are and recognize the important role they play in organizational life. In fact, helping you have a greater understanding of each and learning how they can be effective is the purpose for the writing of this book.

Think about where a building project begins. Everything starts with the blueprint. Our constructing a successful organization requires our developing a blueprint that will produce what we're looking for. So what does it look like? Take a look at the blueprint for becoming an effective organization.

Blueprint For Becoming An Effective Organization

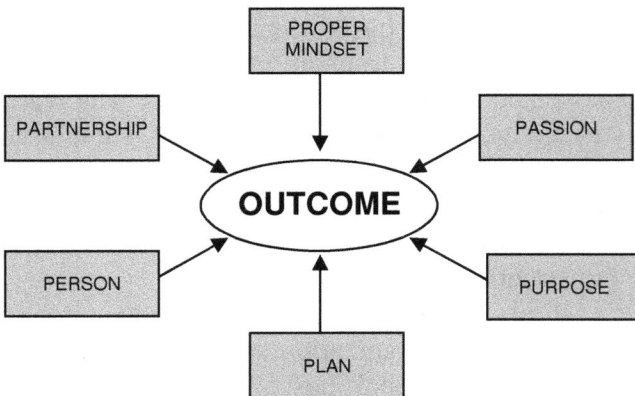

The foundational posts for organizational construction are illustrated in the above diagram. To better understand them, it is important to define each. They reveal that those who follow the blueprint:

- have a **proper mind-set** concerning the characteristics of a healthy organization;
- fulfill the work of the organization with **passion**;
- understand their **purpose**;
- develop a **plan** to accomplish their purpose;
- train each **person** to be effective in their role; and
- have leaders who **partner** together with others as a team.

All must be present for success. As mentioned above, if one is missing the organization comes crashing down.

How healthy is your organization? If it is unhealthy, what are you going to do about it? Will you do what is necessary to turn things around and to accomplish something great once again? Why don't you treat this as a fresh start.

My wife said something very wise to me in the midst of my despair over the health of our organization. She said, "Tim, you're the leadership expert who has the degrees in leadership and administration. You're the one who consults with other organizations about their challenges. What would you tell us to do if you were an outsider consulting us?" Then she said, "Why don't you fire yourself and rehire yourself and act like you're the new leader brought in to lead us. Maybe that would help you know what to do. If you know what you would do in that situation, why don't you just do it!" These were some amazing words that inspired me. Let me share them with you – "Why don't you fire yourself as the leader and rehire yourself right now as if you were the new leader brought into your organization. What would you do as the new leader? Why don't you just do it!"

This book is to help you decode what is going on in your organization and to help you as you make decisions moving forward. I really believe it can help you bring about positive change. I've heard it said before – "Your organization is perfectly organized to achieve the results you're getting right now." If you want to change the results, you have to be willing to make changes to achieve those results.

Do you have your construction clothes on? Are you ready to go to work? Join me on this journey and discover how to lead in a way that matters.

1

THE PROPER MIND-SET

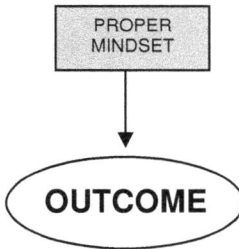

```
┌─────────────┐
│   PROPER    │
│   MINDSET   │
└─────────────┘
       │
       ▼
  ╭───────────╮
 (  OUTCOME   )
  ╰───────────╯
```

---

*H*ow do you define a healthy organization? What is your mind-set on how to measure its health? If we would ask several leaders these questions, they would undoubtedly express many different ideas. Our responses to these questions are crucial because our beliefs about organizational health have a direct bearing on how we perform. Why? Leaders tend to conduct activities that help them meet their definition of organizational health.

We live in a society where the word *health* is used quite frequently. The term is applied to the condition of many things, and one of the best examples is our body. Scientists are working diligently to discover vaccines for diseases. New nutritional programs are being offered at a rapid pace, making promises of fast weight loss or higher energy.

There are certain things that we know about the health of the physical body. We know the parts that make it up must function properly and that it must maintain the right balance for it to experience health. For example, it begins to break down if its blood flow is restricted or too few cells are being produced.

You might be wondering what this has to do with organizational health. Let's define some terms to help us see the connection. First, we focus on the word *organization.* An organization is defined as "a body of people formed to produce a defined purpose." Second, we focus on the word *health,* which means "someone or something that is free from disease." A body that is free from disease functions in the proper manner. We combine these definitions and learn that a healthy organization is defined as "a body of people free from disease which properly produces its defined purpose." Is your organization free from disease?

Just as the physical body operates properly because of the continuing function of its parts and the balance it maintains, the organization operates properly because those who make it up fulfill their functions and the elements of the body maintain their balance. Organizational health is dependent upon its parts working well together. This requires us to have a concern for how well others in the organization are functioning. As the body needs exercise to stay in shape, we need to exercise by investing in others to help them live up to their potential. We need each other. Everyone needs to have the proper mind-set!

THE PROPER MIND-SET:
WE ARE IN THIS TOGETHER

We are unique. We have different talents and abilities which allow us to accomplish different roles. Our health depends on each part of the body performing their roles. Just as the physical body can't operate well without certain body parts, we need everyone involved to maintain a good condition. When someone slacks off, it can have a great effect on overall performance. Our doing our part changes how we feel about ourselves. In fact, we can begin enjoying the work experience when we understand the important role we play.

Philip Yancey and Dr. Paul Brand addressed the health of organizational life in their writings. They compared an organization to the cells in the human body. Dr. Brand wrote, "I sometimes think of the human body as a community, and then of its individual cells such as the white cell. The cell is the basic unit of an organism; it can live for itself, or it can help form and sustain the larger organism."[1] He went on to say that a cell "can be part of the body as a loyalist, or it can cling to its own life. Some cells do choose to live in the body, sharing its benefits while maintaining complete independence - they become parasites or cancer cells."[2]

With this teaching, Dr. Brand showed that we can live for ourselves or for others and that our decision affects the body. Cells that do not live for other cells in the body become cancer cells, and cancer cells cause the body to become unhealthy. The same is true in organizational life. If we begin living for ourselves and not for others, we become cancerous and cause the organization to become unhealthy.

There are four types of cells that affect organizational health. First, there are *cancerous cells*. These are people who are self-centered and have no regard for others. They use others for their own benefit and make no effort to help them live up to their potential.

Second, there are *pre-cancerous cells*. These are people who are moving in the wrong direction. They are being influenced by cancerous cells. The number and quality of actions they perform to help others is diminishing.

Third, there are *pre-functional cells*. These are people who have begun thinking about others more often and perform acts of service in order to help them in their development.

Finally, there are *functional cells*. These are people who are servant leaders. They make sacrifices for the well-being of others and do what they can to help them fulfill their purpose.

Our goal should be to develop an organization made up of functional cells. When this occurs, it affects how we feel about the organization. Dr. Brand emphasized the impact that cells have on the community atmosphere of the body.

In exchange for its self-sacrifice, the individual cell can share in what I call the ecstasy of community. No scientist can yet measure how a sense of security or pleasure is communicated to the cells of the body, but individual cells certainly participate in our emotional reactions.... If you look for a pleasure nerve in the human body, you will come away disappointed; there is none. There are nerves for pain and cold and heat and touch, but no nerve gives a sensation of pleasure. Pleasure appears as a byproduct of cooperation by many cells.[3]

Dr. Brand used the cells in the body to illustrate that we actually experience pleasure when we fulfill our roles in community. Pleasure in the body occurs as a byproduct of corporation and service. The same is true in organizations. Our work with others is to be a pleasurable experience. To ensure this outcome, a proper balance must be maintained. Everyone needs to have the proper mind-set!

THE PROPER MIND-SET: BALANCE MATTERS

Proper organizational health is also dependent upon balance. Think of it this way. Diet and nutrition are hot topics in our modern-day culture. Experts in nutrition advise us to have a balanced diet. Individuals who maintain this type of diet consume vitamins and nutrients that are needed to ward off harmful diseases and illnesses. Organizations are also in a position to experience good health when it maintains proper balance.

The balanced organization does not focus on one area alone. Let's consider an organization that has three areas of importance. They include Sales, Customer Service, and Cooperation (Team Building). This is illustrated in the diagram below. Unfortunately, many organizations lose their balance between areas of importance and become ill. For example, an organization that concentrates on sales alone without a focus on customer service finds itself in a difficult position. Actually, good customer service leads to satisfied customers who then become your best sales people. If the majority of the attention is placed on gaining customers rather than satisfying them once they have been gained, then the organization becomes out of balance and disease enters. The pie chart reveals a balanced approach.

An Organization Pie Chart

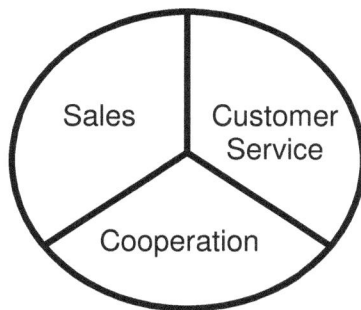

Sales

Customer Service

Cooperation

Let's look at the three divisions more closely. The first part of the pie chart is called "Sales." This relates to people who have not yet associated with your organization. Selling is an important function. Although this may seem to relate only to "for profit" organizations, it does not. Non-profit organizations have the responsibility to convince people that what they offer will meet a need. They are selling them on their service. Without focusing on sales, new customers or clients would not be gained and an organization could not grow.

The second part is called "Customer Service." This relates to all activities that are utilized to meet the needs of clients who have associated with the organization. Again, this is true of both "for profit" and "not-for-profit" organizations. They will either experience satisfaction or will be unsatisfied. The degree to which they have these feelings affects the positive or negative influence they have on potential customers or clients who could be reached through their testimony about their experience.

Customer service is a very important part of the effectiveness of non-profit organizations as well. In a religious setting, it is important for leaders to maintain relationships with those who become members within the congregation. They are the customer who needs effective continual service. Friction occurs often within religious congregations when these relationships are not prioritized. In looking at our challenges, this was a major factor affecting us. Because of re-organization, we lost effectiveness in maintaining these relationships. The result was a lack of commitment and involvement from membership. Customer service needed to be improved.

The third part is called "Cooperation." This relates to the people who are partners in your organization. They determine the success of what you do. Cooperation emphasizes the attention that partners give to one another. Our level of cooperation affects the way we feel about the team. It also affects the results that we produce. Better cooperation leads to more success in both sales and customer service.

Maintaining balance became a weakness for our organization. Rather than being one made up of people who continually helped one another become successful, we became more territorial in our approach, only putting attention on our areas of responsibility. This was not a deliberate act but one that occurred because of my failure in leading our organizational partners to get involved with others on the team. I lost the proper mind-set that "balance matters" by not giving enough attention to building cooperation among our team.

All three parts of the pie affect one another. Each area is critical to the success of the organization, yet often one or more is devalued through the amount of time and energy put forth to ensure its success. If one area is not given the needed attention, it then has a cancerous effect on other critical functions. Imbalance can cause a negative shift in attitude among people within the organization which makes it much more difficult to reach our goals. Why do we not put forth more effort? Everyone needs to have the proper mind-set!

THE PROPER MIND-SET: PEOPLE MATTER

Notice that sales, customer service, and cooperation each identify a group of people. There are those who are not yet affiliated with your organization (sales efforts), are affiliated with your organization (customer service efforts), and are apart of the team (cooperation efforts). Our focus should not be on our efforts but on those we are reaching. Organizations can begin to fail because they become more concerned about what they are doing than who they are doing it for.

The more concerned we become about the group we are reaching, the more effort we will put forth to do what is necessary to reach them, whether it is with a product or a service. There is a progression that takes place which is illustrated below.

The Power of Concern

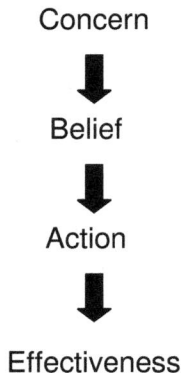

Concern

⬇

Belief

⬇

Action

⬇

Effectiveness

The Proper Mindset

Our "concern" for the individual is the motivating force that influences us to get involved. Our concern is followed by the "belief" that what we offer will improve their lives. Our concern and belief encourage us to take "action" by providing the service or product that we believe will make a difference. Our taking action by providing what they truly need leads us to be "effective" in what we do. If we have no concern, it changes the attitude with which we approach our task. If we don't believe what we provide will make a difference, then it will reduce our motivation to take action. If we don't take action or we take action by providing the wrong product or service, we become ineffective.

Not only should we care about the needs we are meeting through our product or service, we also need to be concerned about the needs of those on the organization's team. Unfortunately, it is easy to forget about people and lose concern for those who are working by our side.

I'm reminded of an event that took place in the early 1900s told by Dennis Perkins. It involved two explorers, Earnest Shackleton and Vilhjalmur Stefansson. Shackleton led a daring expedition in 1914 to reach Antarctica and a year earlier Stefansson led an expedition headed in the opposite direction toward the North Pole. Both ships carrying the explorers and their crews found themselves in great danger. On the way to their destinations they became surrounded by solid ice packs that would be the circumstance that would lead to a difficult struggle for survival.

The outcomes were much different. In the north, Stefansson's crew degenerated into a band of selfish, mean-spirited, cut-throat individuals ending in the death of all 11 crew members. In the south, Shackleton's crew faced the same problems – cold, food shortages, stress, and anxiety –

but his crew responded with teamwork, self-sacrifice, and astonishing good cheer.

In the end, each leader stayed true to his core leadership values. Stefansson valued success above caring for people. He was concerned only for self and consistently communicated his ultimate objective: getting to the North Pole. In Stefansson's words this meant "that even the lives of the (crew) are secondary to the accomplishment of the work!" To the very end, Stefansson denied that his drive for success led to a tragedy for himself and his crew.

In sharp contrast, Shackleton's leadership focused on the value of the dignity of his teammates. At one of the lowest points of his trip, Shackleton wrote, "The task was now to secure the safety of the party." The well-being of his team drove him to put others first. He even gave away his mittens and boots and volunteered for the longest night watches. By valuing each person, he forged a team that was willing to share their rations with each other, even on the brink of starvation. Through his example of sacrificial leadership, he was able to accomplish his ultimate objective: saving the lives of his crew members.[4]

One was selfish and the other was concerned about the needs of others. Shackleton was accepted and appreciated as a courageous leader who put others first and Stefansson was seen as mean spirited and misguided by his ego. The way we think about our role in leadership makes a difference. What do people think about you? Everyone needs to have the proper mind-set!

THE PROPER MIND-SET:
WE MUST BEGIN WITH THE END IN MIND

Many diets and exercise programs begin because individuals look at themselves in the mirror and are not satisfied with what they see. The more unhappy they are, the greater likelihood they will begin adjusting their behaviors to change how they feel and what they see.

Several years ago, just prior to taking the position at Woodland, I had an experience like this. I had gained a lot of weight and was the heaviest I had ever been. It snuck up on me. I was in the habit of getting up and looking at myself in the mirror everyday, just like most do. When you look at yourself each day, it's difficult to notice the gradual change that is happening in your appearance.

Things reached a critical level for me one day when I put on a pair of jeans. It was a pair I hadn't worn in a while. You can imagine what happened. They were too small! I had never had a problem with them before, but now I couldn't get the button to snap. If that weren't enough, I caught a glimpse of myself in the mirror in the jeans. My love handles were all over the place and I was disgusted by what I saw. That's all it took. I couldn't stand the vision of myself and had to do something to change my condition. The diet began and 30 pounds later the jeans went back on.

Organizations need events like this when there has been gradual unnoticed change that has led to poor health. Being shocked into reality by a poor performance report or the loss of a long time client who no longer is satisfied with the quality of our work can be a great thing. It's great if it leads to positive change. When we combine this hard reality with a vision of what we want to become, the diet plan begins that can lead to good organizational health once again. This vision of what we want to become is the end result we're after.

We are to begin with the end in mind when it comes to organizational life. A good example of this is the strategy developed by *Toys 'R' Us* for the Christmas season. The end result they are after is a big vision. It's to share their products with the world. Due to the economic downturn and tough competition, the toy chain has started to change their strategy. Specifically, in 2010 *Toys 'R' Us* plans to "invade" shopping malls around the country by opening 600 express stores and hiring 10,000 part-time seasonal employees. Rather than waiting for customers to come to their traditional stores, they are finding ways to reach out. It's all part of their big vision to share their products with the world. According to CEO Gerald Storch, "We've been very aggressive (about reaching people) during the economic downturn, and this is another aggressive action." They're not asking, "How can we survive this year?" Instead, they're trying to make an impact by asking, "How big can we make this?" It's an impressive display of creative and visionary thinking about how to reach people where they are. It all began by an end result that spurred them into action.[5]

To achieve our end result, success must be defined. What defines success for your organization? That's the end result that you should desire. Your purpose is to achieve whatever that may be. If this end result is not defined, you will not know when you have become successful. You might be saying, "I'm not sure what our organization should look like!" We have already described some elements of a healthy organization through what we've learned about having a proper mindset.

The end result should be that…

- We are a team.
- We are balanced.
- We focus on the people that we are reaching.

Is your organization doing these three things? Although each organization can accomplish these, they may look very different dependent upon their makeup. The role of a leader is to prepare the people by painting a picture of what this success looks like. Once this has been accomplished, the work can begin. So do you have the proper mindsets? Take some time to answer the following questions to better understand the health of your organization.

Questions

- How do people define success in your organization?

- Do people within your organization care about the success of other partners who make it up? Why do you feel this way?

- As you think about people in your organization compared to cells in the body (cancerous, pre-cancerous, pre-functional, functional), which best describes you? Why?

- What does your organization do to encourage serving one another within the organization?

- Does your organization have a balanced approach, putting enough effort toward each organizational focus such as sales, customer service, and building cooperation? Why do you feel this way?

- Use the circle below to draw a pie chart using the focuses of your organization (example: sales, customer service, and building cooperation) as to the amount of attention or emphasis each receives. Under the circle, explain what you could do to be more balanced in your approach.

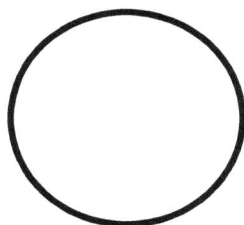

Notes
(Other thoughts related to what you have learned in this chapter)

2

THE PASSION

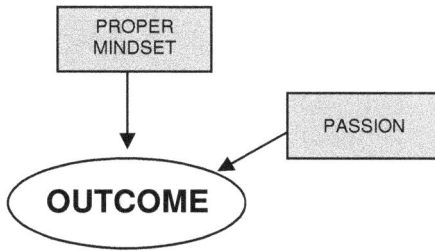

```
┌─────────────┐
│   PROPER    │
│   MINDSET   │
└─────────────┘
        │                    ┌─────────────┐
        │                    │   PASSION   │
        ▼                    └─────────────┘
   ╱─────────╲                    ╱
  │  OUTCOME  │◄────────────────
   ╲─────────╱
```

Our passions are powerful. When they are the same as those possessed by others, they unite us and motivate us to do great things together. They are also the source of motivation that keeps us going when things are tough. They encourage us to get up and try again so that what we are passionate about becomes a reality.

What passions have directed your journey? From the time I received my driver's license, I wanted to learn how to ride a motorcycle. As the years went by this desire began to grow and grow. I would be driving down the road and hear the rumble of a bike come by me. This caused me to dream about what it would be like to feel the wind through my hair and the engine beneath me. If you saw my hair now (gone) you'd understand why these memories are so strong.

One of the obstacles that kept me from fulfilling my dream was my occupation. Because of my work schedule, I was unable to attend the classes I needed. About 6 years ago, at Christmas, I told my wife that I wanted to see if there was some way I could make it happen. I called a local training school to see what they could do. After hearing the challenge I was facing, they decided to get creative and found a way to make my dream come true.

I was pumped. I was so pumped that I immediately bought my first bike, a big cruiser. This is not recommended for new riders because of the size of the bike and the more advanced abilities that are needed. But, of course, I didn't think I was like most riders. That is until the second day I had the bike. I was riding on a very curvy road and was having a blast when I came up to a stop light that changed quickly. No big deal! After all, I had learned how to stop in an emergency through motorcycle school. I used the right breaking technique and thought everything was cool, when all of the sudden the bike started leaning. Over and over it went until it came crashing down. What I had not anticipated was the angle of the road I was on. It was sloped to the right. The tendency of a rider is to make the bike even with the terrain of the road. This is the wrong tactic. The goal is to keep the bike straight up so that it stays in balance.

I was really bummed about the event and felt like a failure. After all, it was day two and I had already dropped the bike. It would have been easy to get discouraged and say, "I can't do this!" or "I just wasted all of that money on motorcycle school and I'm a loser!" Well, at the time, I really wasn't very good and I had lost the battle with the bike. There is one thing that made all of the difference. I still liked the way the wind felt across my face, the sound of the engine, and the free feeling I had riding down the road. It was so awesome that it became my passion.

My failure didn't defeat me, it motivated me. I was determined to learn how to overcome what had made me fall down so I could achieve what I had such a great desire to experience. That's the power of our passion. Unfortunately, we often let our failures deflate us rather than motivate us to be better so that we can obtain the object of our passion.

MOVEMENT IN THE
RIGHT DIRECTION

Just think of what can happen when multiple people who have the same passions work together toward their common goal. Great momentum is created and a sense of enthusiasm begins to drive us. This is the type of organization we're after.

Erwin McManus, an organizational growth expert, wrote about the potential of working together. He brought to mind what would happen if we stop thinking about our organizations as institutions and more as movements. He taught that institutions are only a group of people connected by programs and rules while a movement is a group of people who want to work together to bring positive change in our world.[1] Unfortunately, our organizations often become institutions where we feel confined and controlled rather than empowered to make a difference.

To better grasp this concept of movements and institutions, let's take a look at some marked differences between the two.

Institution vs. Movement

Institution:	Movement:
Program Driven	Purpose Driven
Rules	Creativity
Stagnant	Action
Individual Focus	Team Focused
Unbelief	Belief
Maintenance	Growth

Let's look at each of these comparisons in more detail.

Institutions are program driven and Movements are purpose driven

Organizations that are driven by programs are stuck doing things the way they have always done them. They forget that at one time many of those activities or policies were put into place for a specific purpose to address an important concern. Now, rather than the concern driving what they do, the action or policy becomes the motivating factor. Organizations that fall into this trap have a common belief - "if it doesn't work, we need to do it better!" Some would say – "if it doesn't work, we need to work at it harder!" It may be that if it doesn't work, it needs to die and go away.

Institutions focus on rules and Movements focus on creativity

We also see that rules are compared to creativity. Rules confine us to doing things in certain ways. Obviously, rules aren't necessarily bad. The problem comes when we live for the rules rather than live to accomplish our mission. Living to accomplish mission promotes creativity and new ideas of how we can achieve the end result we're after. This leads into the next comparison.

Institutions are stagnant and Movements take action

Institutions are stagnant, being stuck in a rut, and movements are going places. The key word is "advancement." Movements are making new advancements and are staying on the cutting edge. These productive advancements breathe new life into the organization.

Institutions focus on the individual and Movements focus on the team

Some organizations are focused on the individual rather than the team. In these institutions, there is a protectionist attitude. We are looking out for ourselves because the true goal is not to dream bigger dreams for the company (it's not encouraged), but to keep our job and keep bringing in the dough. Movements encourage people to dream and to share ideas with each other. They believe in the power of Collective IQ which promotes the understanding that we are smarter together than we are apart. One person's ideas spark new thoughts in others which in turn lead to discussion resulting in stronger ideas.

Institutions don't believe and Movements do believe

We also notice that an institution is made up of people who don't believe that the product or service is making a positive impact. This is compared to a movement who believes in what they do. If we don't believe in what we provide, it is difficult to have a positive attitude. This connects to our passion. The more we believe in our work, the more passionate we become, and the better our attitude. This is illustrated below.

Attitude Meter

indifference

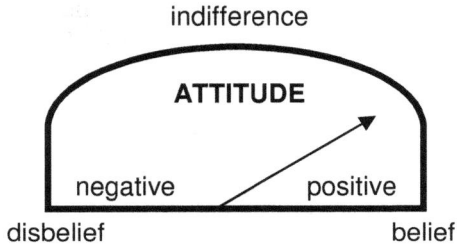

The goal is to help people see the importance of what the organization is established to accomplish and to stoke their passion for it. Some experience disbelief, not believing in the mission of the organization. Others are indifferent. Finally, some are stoked about fulfilling the objectives because of the value they place on organizational efforts. The level of belief is directly connected to the condition of our attitude.

You can easily identify bad attitudes by noticing three things: the quality of a person's work, the amount of time they are willing to give to do their work, and the words they are using. Again, these often become issues because of their feelings about the importance of the product or service they provide. Becoming a movement begins with the belief that our efforts matter. We ask ourselves the question – "Is it worth it?" If it is, we become committed to the cause and give our best effort and move toward using what we provide to make a difference.

Institutions are concerned about maintenance and Movements are concerned about growth

There is one final comparison. Institutions want to maintain what they are doing and movements want to grow. We often think of growth only related to expanding the organization to bring in more profit. Although this can be an important goal, the more important growth happens when we help to develop people within the organization. We want to help them become better at what they do and become better in character as they learn to sacrifice for others in the organization. Focusing on maturing people leads to a feeling of personal significance which leads to a positive feeling about who we are as people working together.

IT'S US NOT ME

To have the right passion, we need to have the right belief about our involvement in organizational life. What should we believe? The success of the organization is not about me, it's about us. We can tell if this is our attitude by looking at the "A, B, C" principle = If "A" and "B" happen, then "C" will occur. If it is only about me, I will say - "If A and B happen, I WILL get C" (whatever it is that I desire). If it is about us, I will say - "If A and B happen, WE WILL get C" (whatever it is that we desire).

"A" and "B" represent what we do to achieve our desires. We can tell if they are coming from a passion for organizational health by determining whether or not those actions will have a positive or negative affect. Those who are more passionate about self tend to perform actions that are detrimental toward an atmosphere of cooperation.

You may disagree with this statement. After all, we know that people, who are motivated by money and not by the fulfillment of a greater purpose, may be very effective in sales which results in a positive affect on the organization. Their concern, however, is not about organizational success but in personal gain. This attitude works against a sense of teamwork which would lead them to put forth effort to help others become more effective. Wouldn't the organization experience a greater increase in sales if the person began working with others to help them improve their abilities? The answer is "yes!"

Why don't we help others? Often we encourage a competitive environment which pits one person against the other. Unhealthy competition in the organization can lead to manipulation and animosity. We know that our passions are off if we are more concerned about "winning" than "helping." This "winning" attitude causes unhealthy behavior. It destroys the sense of team and reduces the power of our working together and synergy disappears. Unfortunately, competition within the organization can keep us from being competitive outside of the organization.

OUR MOTIVATION

Our passion and motivation go together. Whatever motivates us becomes the passion of our lives. All of us are looking for happiness. People attempt to achieve this through three specific means - power, pleasure, and people.

Power

Some try to find it through their power. In other words, they believe they will experience happiness when they have authority. Having authority is not necessarily a bad thing. It becomes unhealthy when our achieving authority is the primary motivation of life. Desiring a position of great influence can be a worthy goal, especially when we want to increase our influence for a positive purpose. Problems come when our passion is not to influence but to receiving recognition. We mistakenly believe that our receiving it will bring us the fulfillment we're looking for.

Pleasure

Others try to find their happiness through pleasure. Often people attempt to achieve this through their ability to purchase items that will give them what they are looking for. Purchasing items to have pleasurable experiences is not necessarily a bad thing. Many items that we attain become exciting additions to life that become great stress relievers. Once again, the difficulty comes when we live to buy these things and believe that they will provide the happiness that we're looking for.

Materialism has become more and more of a motivating factor over the years as new products and services have become available to us. This is especially seen through what college students are saying is "very important" or "essential" to them? Data from a Higher Education Research Institute Study of nearly 220,000 first-time and full-time freshmen revealed that 78.1 percent wanted to be "well off financially." This compared to only 42.2 percent of college freshmen in 1969. Another surprising statistic is that only 48 percent wanted to develop a "meaningful philosophy in life," compared to a whopping 84.9 percent in 1969.[2] Things really are changing.

I know what falling for the pleasure trap is like. It's time for full disclosure. I have been diagnosed with Bipolar Type II disorder and have suffered with this disease for many years. Around the time of my diagnosis, I was surprised to discover that many leaders of large corporations also live with this challenge. Although a large part of the disorder is seen through major mood swings from high to low, there are two other characteristics about this condition that are common. One is that we tend to take risks, believing that everything will always work out. This can actually be a great quality. The reason why so many who have the disorder have become successful is because of their willing to take great risks. Although it can have positive effects, it can also have very devastating outcomes. Often, we believe things that aren't very realistic and make quick decisions that are unwise which lead to failure.

The second common characteristic is living for pleasure. Many who suffer from this disorder have a problem with buying things. Once again, they believe that their finances will work out allowing them to make these purchases. Typically, they are big ticket items.

In my case, I would buy boats. You might not think that's a big deal. You haven't heard the whole story. I bought six boats. There were moments that I had more than one at a time. I obsessed about them, believing I needed them. I wanted something as an escape to get my mind off of the stresses of my life in other areas. In the end, it only added more stress as I over-extended myself and was overcome by major buyer's remorse. I lived for those things and they didn't really satisfy.

Work became a way for me to get what I wanted. In the process, I hurt my family by putting us in financial jeopardy to satisfy my desire for pleasure. Believe me, in the end it doesn't work. Just as a side note, if you'd like to hear more about my story and practical ways to deal with emotional issues, I encourage you to read my book entitled *One Fry Short: A Journey Toward Self Discovery and Emotional Success.*

Personal Relationships

The third way we attempt to experience happiness is through our personal relationships with others. This last motivation is the real source of happiness and joy. It's ironic that if we attempt to find happiness through power or pleasure and do not, we often act out against the one thing that does. We lash out at those around us.

Many become angry and bitter at others because they believe they are standing in the way of their happiness. For example, if someone is in a position of authority and feels threatened by another, the person living for power will act out aggressively, either passively or openly, to destroy them in order to maintain their position. Let's think about it related to pleasure and our attempting to gain it through what we buy. If someone is standing in the way of our making more money, we will also act out aggressively toward them. Again, we become bitter and angry. This can also happen within an organization if one person becomes more successful and gains more clients than another. The result of this unhealthy competition is strained relationships.

The goal of leadership in an organization should be to influence those within it to concentrate on what matters most. They are to encourage the building of relationships. How do we do this? There is a progression that leads to our experiencing happiness in the work environment. Take a look at the illustration.

The Connection Progression

Connection

⬇

Passion

⬇

Willingness

⬇

Investment

⬇

Unity

⬇

Happiness

The more "connected" we feel to others the more "passionate" we are about them and the more "willing" we are to "invest" in their success. "Unity" is the result. When there is unity there is "happiness." Wouldn't it be awesome to work in a place where everyone is happy? We can dare to dream!

One difficulty is that people in the workplace don't feel appreciated, even by those who are over them. According to a 2010 Cornerstone On-Demand survey, the majority of Americans (54 percent) think their coworkers appreciate them the most at work and not their supervisors. Only 30 percent believed that their supervisors appreciated them more than their peers.[3] If we are going to lead partners to be stronger in relationships with each other, we as leaders must set the example. To address this, we must determine what causes people to feel appreciated.

People feel appreciated when they are...

- ✓ known
- ✓ noticed
- ✓ encouraged
- ✓ helped

All require our time. Each connects specifically to the investment part of the progression. We invest in people when we take the time to get to know their names and who they are. Our interest in them places value on them. We also need to notice them. Our acknowledgment that they are in the room speaks to their value. They are worth our attention. We need to be people who encourage them by doing things that prove our concern for their well being. Finally, we need to help them become better and to live up to their potential. The absence of these things destroys the sense of appreciation. As leaders who desire to make a positive impact we must be passionate about people.

A BASIS FOR CHANGE

A key element is to lead others to be passionate about accomplishing something that is greater than themselves. Passion is the basis for change. We don't change for the sake of change, we change when there is a benefit. We see the purpose for that change and believe in it.

Jim Harrington, Mike Bonen, and James Ferr addressed the topic of creating urgency for change. They explained:

> ... Change is driven when a significant gap exists between a vision of the future that people sincerely desire to achieve and a clear sense that they are not achieving that vision. As this recognition grows, so does their willingness to change their perspective and to try new approaches. This is the point at which they are experiencing creative tension.[4]

This understanding of creating change is also taught by John Kotter, a professor at Harvard, who wrote: "establishing a sense of urgency is crucial in gaining needed cooperation."[5] When there is urgency there is tension that exists because of our need to act now. This tension is a motivating force.

People in organizations are more willing to change when they experience tension because an objective they desire is not achieved. They recognize that their current methods are keeping them from accomplishing their goals, and they believe that change will assist them in their efforts.

For there to be change, there must be dissatisfaction. There are two types of people - the satisfied and the dissatisfied. When I hear the word "dissatisfied," I often think of someone who is negative. In this case, dissatisfaction is positive. Those who are satisfied with the way things are will not make sacrifices to bring change. The difference between a successful and unsuccessful person within an organization often times is dependent upon their dissatisfaction. Typically, those who rise to greater levels of leadership are dissatisfied with the status quo. They are constantly thinking about needed changes that would bring improvement.

If leaders attempt to change an organization without creating a sense of dissatisfaction within those who make it up, they will find themselves in a difficult situation. Leaders are met with great resistance because organizational partners are not ready to willingly put forth the effort to accomplish the goal. They are not willing to make the necessary sacrifices. This brings conflict and an unhealthy tension. Instead of creating "creative tension" we have created "destructive tension."

The solution is to lead people to be dissatisfied. The leader must help them understand the benefit of what will happen if the organization reaches its goals. The more passionate they become about this, the more dissatisfied they will be with what is standing in the way.

KOTTER'S CHANGE PROCESS

Are you ready for some good news? Ineffective organizations can change and become successful. We just need to know what we can do to bring it about. Kotter, an expert in the field, defined a practical eight-stage process to facilitate change in an ineffective organization. We'll take a look at each. Make sure that you notice that the first stage is to "establish a sense of urgency."

Stage One: *Establishing a sense of urgency.* This includes examining the market and competitive realities and identifying and discussing crises, potential crises, or major opportunities.

Stage Two: *Creating the Guiding Coalition.* This includes putting together a group with enough power to lead the change and getting the group to work together like a team.

Stage Three: *Developing a vision and strategy.* This includes creating a vision to help direct the change effort and developing strategies for achieving that vision.

Stage Four: *Communicating the change vision.* This includes using every vehicle possible to constantly communicate the new vision and strategies and having the guiding coalition role model the behavior expected of employees.

Stage Five: *Empowering broad-based action.* This includes getting rid of obstacles, changing systems or structures that undermine the change vision, and encouraging risk-taking and nontraditional ideas, activities, and actions.

Stage Six: *Generating short-term wins.* This includes planning for visible improvements in performance, or "wins," creating those wins, and visibly recognizing and rewarding people who make the wins possible.

Stage Seven: *Consolidating gains and producing more change.* This includes using increased credibility to change all systems, structures, and policies that don't fit together and don't fit the transformation vision. It also includes hiring, promoting, and developing people who can implement that change vision. Finally, it involves reinvigorating the process with new projects, themes, and change agents.

Stage Eight: 8. *Anchoring new approaches in the culture*. This includes creating better performance through customer and productivity oriented behavior, more and better leadership, and more affective management. It also includes articulating the connections between new behaviors and organizational success and developing means to ensure leadership development and succession. [6]

Each of these are critical in bringing change. During the remainder of this book, much of what Kotter emphasized in this change process will be addressed. Before we get into that information, it's important to get a bearing on where we are now. With this change process in mind, along with the other concepts taught in this chapter, answer the following questions to better define your organization and how you can develop passion.

Questions

- Do you feel your organization is operating as an institution or as a movement? Explain.

- Do you feel that those in your organization are motivated by maintaining good relationships or by power and pleasure? Why do you feel this way?

- Do you feel that change is needed in your organization? Why or why not?

- What has the leadership of the organization done to establish a sense of urgency?

- What is the competitive reality between your organization and others?

- What crises, potential crises, or major opportunities is your organization experiencing?

- Is there a group of people guiding the organization as it deals with change? If so, who is it and do they have enough power to lead the change?

- Does this group work together like a team? Why do you believe this?

- Has the leadership communicated a vision for the future for your organization? If so, what is it?

- Has the leadership developed a strategy for achieving that vision? If so, what is it?

- How are leaders in your organization communicating the new vision and strategies?

- Are the members of the leadership group who are responsible for the change good role models related to the behavior expected of those in the organization? Explain.

- Is the leadership group responsible for change working to remove obstacles? How?

- Is the leadership group responsible for change encouraging organizational partners to take risks and develop new ideas, activities, and actions that could assist in the change? If so, describe how they are encouraging this to occur.

- If the leadership group responsible for change has been effective, have they been using their credibility to move forward in creating other needed changes?

Notes
(Other thoughts related to what you have learned in this chapter)

3

The Purpose

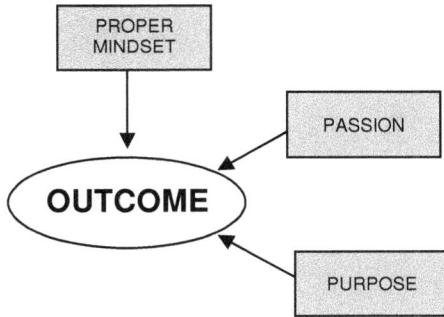

---

*T*he launch of the space shuttle is one of the most amazing sights Florida residents have the opportunity to witness. Several years ago, I turned on the television and learned that the shuttle was about to launch. I hurried outside my home in Bradenton (which is on the other side of the state from Cape Canaveral) to witness this incredible site. I stood in awe as I watched this man-made hunk of metal zoom through the sky toward, what I could only imagine, was an amazing destination.

Just think of what it takes for a shuttle mission to be successful. It's successful because it really is a "mission." There is an ultimate goal that has become a passion which drives the efforts of those involved. Mission is extremely important to organizational life. Although this is true, not all organizations focus around a mission. Let's consider three types of organizations.

TYPES OF ORGANIZATIONS

The "Me" Organization

In this type of organization there is a culture of self-centeredness. People are thinking only about self and aren't willing to make sacrifices. The story of Steffansson's expedition crew shared in Chapter One is a classic case of a "Me" Organization. They were only thinking about themselves and the result was tragic. There was literal death. Organizations find themselves in this condition because of the influence of the leader. Steffansson's attitude encouraged this savage mentality that brought destruction. The same can occur in the organizations of our day. Although we might not turn on each other in a way that we kill one another, we might turn on each other enough to kill the organization.

The "Misguided" Organization

In this type of organization, there is a culture of confusion. The difficulty is not selfishness. In fact, there may be a great concern for people. However, their lack of knowledge and/or ability keeps them from being successful. Leaders don't know how to properly lead by setting the course for the organization.

The "Mission" Organization

In this organization, there is a culture of teamwork. There is a clear understanding of what the organization is to accomplish and everyone is passionate about achieving it. This sense of mission leads to a willingness to make sacrifices and a desire to work with others to fulfill their goals. The expedition led by Shackleton is a fantastic example of a "Mission" Organization. Although the goal was to make it to the pole, the greatest mission was to protect the people on the way to where they were going. The primary goal was to meet their needs, even if it required giving of self. Organizations find themselves in this condition because of the influence of the leader. Shackleton's attitude encouraged a spirit of giving that brought safety and success. Again, the same can occur in the organizations of our day. Those who have a mission founded on clear needs that can be met encourage partners to make sacrifices for this worthwhile goal. The result is not organizational death but organizational life.

Obviously, the most successful of these is the "Mission" Organization. It begins with knowing what we are to accomplish. The first responsibility of leadership is to define the mission. Because of its importance, it might be helpful to consider the key leadership team of the organization as the "Mission Control Team."

THE THREE QUESTIONS

Determining the mission is a big task. To properly accomplish this, we need to consider some critical issues. Three questions should be answered that help us discover the mission we should pursue.

Question One: What needs are present that need to be met?

The first issue is determining what needs are out there. When I say "out there" I mean things that are in our communities or world that need to be addressed. These become the catalyst for our doing what we do. The goal is to meet a need that will improve the quality of life for those who receive our product or service. Determining the need or needs we can meet are important because our meeting of needs is what gives us a sense of purpose. It gives organizational partners value. We are all looking for this.

National Geographic identified three regions of the world in 2005 where people have consistently shown longer life spans. They include Okinaway, Sardinia, and Loma Linda, California. Researcher Dan Buettner decided to do follow-up research and discovered that on the Nicoya Peninsula in Costa Rica it was common for people to live past 90 and even into their 100s. This fascinated him and he decided to take a team to the region to discover what factors caused this. They found that their long life was affected by things that we would expect like diet, their source of water and their exposure to the sun. The study also discovered other factors that were instrumental. They found that a primary cause for their long life is their strong sense of purpose. They feel needed and want to contribute to a greater sense of good.[1] What does this teach us? Our having purpose really does effect how we approach life and how we feel about who we are. We should want people in our organizations to feel good about what we ask them to do.

Question Two: What can we provide as an organization to meet the needs that need to be met?

The second issue is the organization providing what is necessary to meet the need. It might be that there is no product or service available to meet that need. There may be a product or service available that is offered but there may not be enough of it. Also, there may be enough of the product or service but the product or service is lacking in quality and needs to be improved upon. There must be a clear picture of what could be done by the organization to make a difference. To be a "Mission" Organization, people must know there is a purpose for what they do.

Question Three: What are we capable of doing?

The third issue is recognizing our capabilities. It may be that we have a great idea about how to meet the need but the reality is that we aren't capable to do it. We are

---

powerless. Defining our power is important. Our power is made up of our…

- Time
- Abilities
- Resources

We are incapable to do something about a need if we don't have the time, ability, or resources to address it. So what's the answer? It's obvious! Make more time, find people who have the abilities to do it, and find the resources necessary to make it happen. If we don't have these things or the ability to get them, we have no business making meeting the need our business. Understanding this concept should give us clarity about what we can and cannot do as an organization.

MISSION VS. PURPOSE

Once we answer these three questions and know what we are able to do, it's time to carry out the mission. Carrying out that mission becomes our purpose.

There is often confusion about the difference between "mission" and "purpose." Let's define the two:

Mission is the ultimate result we are after.

Purpose is the definition of what we must do to make it to the ultimate destination we are after.

A good definition of the word "purpose" is the word "definition." You're probably wondering, "What does that mean?" The purpose of the organization "defines" what we are to do.

Let me explain it this way. The mission of the space program might be to go to Mars. That is the ultimate result they are after. There are many actions and attitudes that have been defined which must be addressed to accomplish that goal. It is their purpose to address what must be done so that what has been established as the ultimate goal can be accomplished.

Problems occur when "mission" and "purpose" are not both addressed. Some determine the ultimate goal they are after (their mission) without describing what must be done (their purpose). Some determine what should be done (their purpose) without giving a clear definition of the ultimate result they're after (their mission). Can you imagine what would happen if the mission of the space program were not defined? Their purpose would be to develop a rocket that worked really well with pilots who were exceptionally trained but who would have no idea of where to guide the spacecraft. What's the result? Confusion!

In reality, what we are doing is taking us somewhere. In other words, the actions we are performing will have an ultimate result. This brings the truth to mind that *we are perfectly organized to get the results we are currently experiencing.* If we don't like the results, then we must change what we're doing. Everything in the organization must be organized to take us where we want to go. We will do it if the results are worth it.

THE RESULTS OF FULFILLING OUR PURPOSE AND ACCOMPLISHING OUR MISSION

What are the results of fulfilling our purpose and accomplishing our mission? There are several.

Result #1: *An agenda is set for the organization.*

We can set an agenda to get there because we know what we are ultimately after. A reason exists for everything that we do. If we are not accomplishing our mission, then we begin working on solutions to problems which are standing in the way. Knowing our mission reveals problems that exist within an organization. Challenges are exposed and positive changes can be made.

Result #2: *The organization is unified.*

Partners of the organization want to know why they are needed. They want to be an important part of something meaningful. Moral is built and momentum is created when there is a sense of purpose. McManus dealt with the importance of momentum and introduced a formula for it. He wrote:

> The formula for momentum is $P=MV^2$. "P" being momentum, "M" equaling mass, "V" equaling velocity... Mass equals people. Without people there is no momentum. When people move together with common purpose, momentum happens.[2]

The word "velocity" is also important. It refers to the direction in which we go. It is movement toward a destination. When we put all of this together we learn that momentum happens when multiple people come together moving in the same direction. The larger the mass of people moving in the same direction, the greater the momentum. The opposite is also true. If a mass of people begin moving in different directions, momentum is lost. Our moving together in mass toward an important mission creates momentum that can push us through the challenges that may come. I've heard it put this way. We can be a "critical mass" moving in the same direction or a "critical mess" moving in opposite directions. Our being unified leads to the next result.

Result #3: *Teamwork grows*.

Pat MacMillan, an expert in team leadership training, wrote: "a clear, common, compelling task that is important to the individual team partners is the single biggest factor in team success. All the team workshops in the world pale in significance in comparison to a clear and challenging task or goal."[3] A task or goal gives us something with which to align people in an organization.

Automobiles come to mind when I think of the word "alignment." Wheels that are not aligned cause the car to be off course and the driver must exert much energy to keep the car on the road. The solution to this problem is to have an expert perform an alignment.

Leaders must think of themselves as experts who perform alignments. This occurs when there is a link between the goals of organizational team partners and the purpose of the organization itself. Therefore the leader who performs an alignment must link the goals of partners with the defined purpose. The absence of this link causes misalignment which pulls people away from the identified purpose. Workers have goals that are contrary to those of the organization. Competing goals kill teamwork. Unified goals result in our working together and producing multiplied results. This leads to the next result.

Result #4: *Frustration is reduced*.

We need to know what really matters. This helps alleviate things that don't, things that have become frustrations. The goal is to reduce these negative feelings. This is done by adding and subtracting. Guess what that means! Leaders must become great "organizational mathematicians." We add new actions that will help us move toward our goal and we subtract things that do nothing to help us reach our goal. We can blame it on our mission because the mission rules. No longer can we use the mantra – "We've always done it that way!" This is no longer a legitimate reason to do what we do. Rather, we choose to do what works and stop what doesn't.

Let's stick with the alignment illustration that we used in the previous point. Just as we align the goals of people with the goals of the organization, we are to align the actions of the organization with the mission and purpose. Remember, the mission is the ultimate result we're after. The purpose defines what we are to accomplish to get there. Finally, there are the actions that we perform to accomplish our purpose. All three of these are to be in alignment.

As mentioned before, I think about automobiles when I think of the concept of alignment. With this in mind, take a look at the illustration found below.

AN ORGANIZATION IN ALIGNMENT

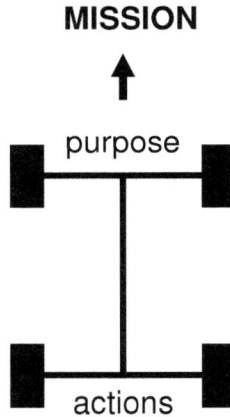

MISSION

↑

purpose

actions

The illustration reveals an organization (like a car) that is moving toward its mission (it's final destination). The purposes established to help it accomplish the mission (represented by the front wheels) are aiming toward the mission. In other words, everything that defines its purpose is directly aimed at the mission. We also notice that the back wheels in the illustration (which represent actions or activities) are aligned with purpose. Our actions are important because they propel us. We can think of the organizational car as one which has rear wheel drive. The propulsion comes from the actions we perform. In this illustration, actions are pushing us toward accomplishing our purpose, which in turn, directs us to our mission. This is an aligned organization. Take a look at the next illustration.

AN ORGANIZATION ALIGNED WITH MISSION BUT MISALIGNED WITH PURPOSE

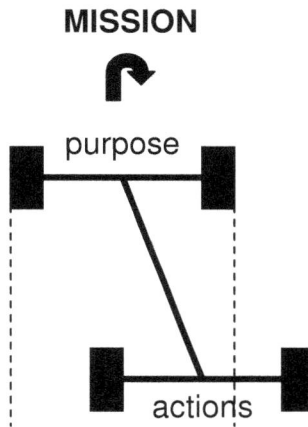

MISSION

purpose

actions

You will notice that the front wheels (representing purpose) are aimed at the mission but the back wheels (representing actions) are not in alignment with the front wheels. Not all of the organization's actions are aligned with purpose. Notice the dotted lines. If the wheels were in alignment, the back wheels would be between the dotted lines. Everything that is not inside the dotted lines are out of alignment. These are areas that must be addressed. The actions are being performed outside of the lines need to be subtracted. Notice the empty space between the dotted lines and the back left wheel. This reveals that there is nothing under that area of purpose which the organization is involved in to accomplish what must be done. Fulfillment of purpose is our goal and is the measurement of success. If we are not accomplishing our goals, change must take place. Since this is the case and there is a gap in our activities related to purpose, then something needs to be added. Remember, leaders are organizational mathematicians. Our adding and subtracting aligns us. What if we don't? Take a look at the arrow in front of the car. If we neglect making changes, the car in this illustration begins to spin out of control.

Let's take a look at another alignment challenge.

AN ORGANIZATION MISALIGNED
WITH MISSION

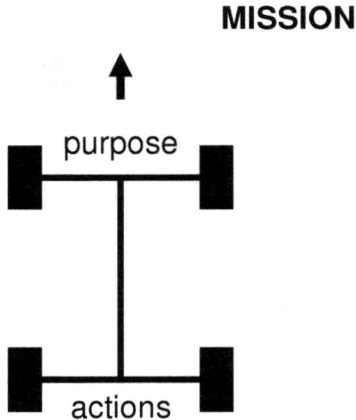

MISSION

purpose

actions

In this case, the front wheels are not aiming toward mission. They are aiming away. This highlights the challenge of an organization that's purposes are not directing it to fulfill its mission. Once again, our mission is the ultimate destination we're aiming for and our purposes define what must be done to get there. It may be that the mission has been established and communicated but leaders within the organization have not bought into its importance. Their purpose is to continue to do what they have always done even though doing so doesn't lead them to the destination that has been determined. The actions of the organization are in alignment with previously established purposes. Again, the problem is that what has always been done doesn't get them to where they need to go. There is a final illustration.

AN ORGANIZATION WITH NO WHERE TO GO

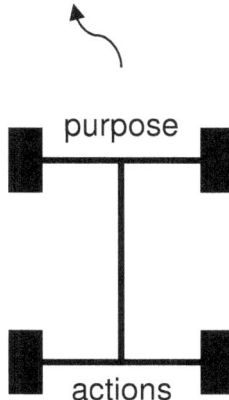

purpose

actions

In this illustration, the word "Mission" is gone. There is no mission. No one has determined the ultimate destination of the organization. Although there is no mission, certain tasks have been established that leaders believe should be accomplished. For whatever reason, they see these actions as important and it becomes their "purpose" to see that they are achieved. As in the previous illustration, it may be that it is what they have always done. They are moving, but they're not certain about the ultimate destination they are to aim for. When this is the case, each leader of the specific areas of the organization decides on what must be accomplished in their group. There is no clear direction of what the divisions are to do together to reach a common destination. What typically occurs is the division of the organization that has the most immediate need becomes the pace setter for a period of time and leads the organization in their direction. Because there is no clear destination for the entire organization, it takes different paths at different times leading to confusion and possibly even death. This is illustrated by the arrow going in multiple directions. It may also find itself going in circles.

This reminds me of an article I read recently about a certain creature that has directional issues. In Uruguay and Argentina there is an amazing occurrence that takes place called an "Ant Death Spiral." Robert Krulwich, in an NPR article, pointed out what leads to this incredible event. The ants in the circle called Labidus Praedator are completely blind. To move in a direction, they sniff the trails that are left behind by the ants that are in front of them. They also leave a trail, allowing other ants to follow. This works really well as long as they move in a straight line in one direction determined by the leader. The problem comes when the ant leader starts to loop and intersects with its old trail. The whole group then turns in on itself and they all become caught in a vicious circle. What is amazing is that the lead ant basically takes over by accident. It's a random act. She follows along and then finds herself in a gap where there is no one to follow. In the confusion she turns, and without meaning to, leads the group and the "Ant Death Spiral" begins.[4]

Isn't that what happens in the organization that has just been illustrated. When there is a gap of leadership and no mission has been established, someone steps in for a while and leads in their own direction. The organization continually changes direction because of this. Eventually, the organization finds itself going in circles that we can call the "Organizational Death Spiral." This is what occurs in the absence of a leader who establishes mission.

A STATEMENT OF MISSION
AND PURPOSE

One of the best ways to communicate our mission and purpose is to state it in such a way that it is motivating to organizational partners. It is vitally important for the Mission Control Team to create a mission statement that clearly presents the ultimate goal that we are after as well as what must be done to see it come about. An effective mission and purpose statement meets some specific criteria. It must be clear, compelling, and measurable and should answer three important questions for the individual:

- Do I understand the mission and purpose?
- Are they worthy of my sacrifice?
- Am I willing to be held accountable for reaching them?

The Mission Control Team must create a purpose statement that encourages people to say yes to all three questions. Our mission and purpose should be worthy of our conviction, passion, and enthusiasm.

How can we state our mission and purpose? We can do it by filling in the blanks. How would you fill in the following blanks?

To _____

_____

_____

(our mission – the ultimate goal)

As we _____

_____

_____

(our purpose – the definition of what we must do)

The first blank begins with "To..." and connects to what we are to ultimately do? What is it? Again, remember the criteria. It meets a need that we have the ability to meet through the resources we posses.

The second blank begins with "As we..." As we do what? Whatever it is helps us accomplish what our mission says we are to do.

EACH PROBLEM REQUIRES THE ATTENTION OF MISSION CONTROL LEADERS

Some may take issue with my referring to "problems." Those who feel this way may prefer that they be called "challenges." There is a reason I refer to them as "problems." Problems are meant to be solved! I love how things connect together. We are organizational mathematicians and we find the problems that need to be solved. We've already learned about one problem. It's the alignment problem. We must add and subtract to correct it. It sounds so easy, but the truth is that many don't do it. Why? There are two other problems which must be solved.

- The Commitment Problem: a lack of commitment from the leadership toward accomplishing the purpose. They don't believe in the need. They aren't passionate about it. Because they have wrong passions, they are unwilling to change.
- The Communication Problem: poor communication about the purpose. They don't understand what they are to do and why they are to do it.

Both are problems which must be solved. Let's deal with the "Commitment Problem" first. Those who have it lack a critical attitude that is needed in an organization – the "whatever it takes attitude." Misaligned members of organizations have the wrong mentality and make a mess of things because of their unwillingness to sacrifice. They are unable to fulfill the mission because their attitude gets in the way. I keep coming back to this, but think again about something we have learned. We can't fulfill the need which is our mission if we are unable to do it. Bad attitudes affect our ability. There is nothing more destructive in organizational life than a bad attitude. You can easily tell when they are present. It is often seen in the quality of work and an unwillingness to go the extra mile. What's the answer? Do what you can to change the attitude or find people who have the right one.

There is a second problem, the "Communication Problem." Leaders must make certain that the purpose is before the people in the organization in clear and creative ways. People easily forget about purpose because of the vast amount of information they take in concerning areas of life other than work or organizational involvement. The more we think about something, the more weight we put on it. If we don't think about purpose and only concentrate on a task, we lose our enthusiasm and become burdened by what we do rather than empowered by what we do. We become complacent rather than inspired. Partners tend to drift off course if the destination is not regularly before them. A leader can't let them forget!

We have learned the importance of having a purpose and a Mission Control Team that works toward its completion. Are you committed to be a part of such a team? Answer the questions below to help you as you continue to grow and make important changes in organizational life.

Questions

• Who in your organization is responsible for developing the mission?

- What research has been done to ensure that this mission is appropriate?

- What is the purpose of your organization? If there is a purpose statement, write it out.

- Are you aligned with the mission of your organization? Why or why not?

- Do you have a passion to fulfill the mission of the organization? Why or why not?

- Does this purpose clearly set the agenda for the organization in such a way that goals can be defined? Why or why not?

- Do you believe the purpose of the organization is worthy of your time and effort? Why or why not?

- Are you willing to be held accountable for the achievement of this purpose? Why or why not?

- Are the activities performed by the organization in alignment with the purpose? Why or why not?

- Are the partners of the organization in alignment with the purpose? Why or why not?

- Has the organizational purpose led to a greater sense of unity among partners of the organization? Why or why not?

- Do organizational partners cooperate with one another in achieving the purpose? Why or why not?

- Does the purpose of the organization provide a good means by which to measure success? Why or why not?

- Where is your organization weak in relationship to achieving purpose?

- What are the possible solutions for overcoming these weaknesses?

Notes
(Other thoughts related to what you have learned in this chapter)

4

The Plan

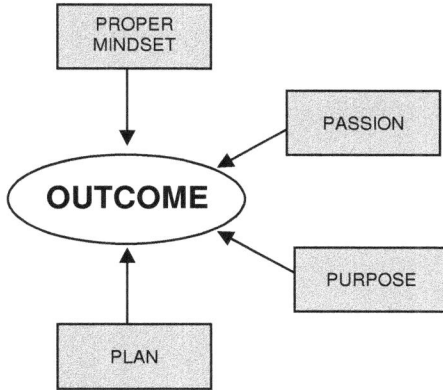

---

*O*ne of the most important roles of a leader is to develop a strategy that helps the organization reach its goals. Peter Senge, in his book *The Fifth Discipline,* tells the reader to imagine that their organization is an ocean liner and that they are "the leader." He writes,

> What is your role? I have asked this question of groups of managers many times. The most common answer, not surprisingly, is the captain. Others say, "The navigator, setting the direction." Still others say, "The helmsman, actually controlling the direction," or, "The engineer down there stoking the fire, providing energy," or, "The social director, making sure everybody's enrolled, involved, and communicating." While these are legitimate leadership roles, there is another that, in many ways, eclipses them all in importance. Yet, rarely does anyone think of it. The neglected leadership role is the designer of the ship. No one has a more sweeping influence than the designer. What good does it do for the captain to say, 'Turn starboard 30 degrees when the designer has built a rudder that will turn only to port, or that takes 6 hours to turn to starboard? It's fruitless to be the leader in an organization that is poorly designed. Isn't it interesting that so few managers think of the ship's designer when they think of the leader's role?[1]

Just as a designer draws plan for a ship or an architect draws plans for a building, we are to determine a plan that is used to help construct the organization.

A PLAN FOR MOVING PEOPLE

Contractors should not build without a design plan. Neither should organizational leaders who are attempting to develop something that is functional and successful. The health of the organization is dependent upon the creation of a plan to accomplish its mission. As we learned in Chapter Three, the mission control team has the responsibility to produce a plan that will accomplish this end result.

In this chapter, we'll learn how to develop a plan that produces what we're looking for. To begin, it is important to define a term. What is a "plan?" A "plan" is defined as "a scheme for making, doing, or arranging something..."[2] The key word is "scheme." This word reflects what we must accomplish in order to produce what the organization has been established to produce. Another word that expresses this idea of a "scheme" is the word "strategy." We need a strategy that we can follow that helps us arrange our tasks in order to reach our goals.

This book began by comparing organizations to a beach front home. With that in mind, let's think about our strategy as a construction plan for our organizational beach front home. The design provided in this chapter is a simple construction plan. The purpose of this plan is to help us produce people who are connected to the organization, believe in what it is achieving, and are involved in moving the organization toward success.

In order to develop this plan, it is important to define more terms that are critical components for construction. Each term emphasizes a group of people associated with the organization.

- The Contact = a person or organization who is a potential client. People in this group are in need of a product or service you provide that can be of benefit.

- The Customer = a person or organization who purchases your goods or services. A customer is someone who has made a commitment to acquire these goods and services due to a belief that what is offered will meet a need and provide a benefit.

- The Committed = a person or organization who believes in the product or service which has been acquired to the point that they are willing to share information about the product or service with others.

The goal of the organization follows what I call a "baton approach." As a runner passes a baton in a relay race, we pass people from one leader to another as they become more connected and involved with the organization. Specifically, the desire should be to lead a Contact to become a Customer, and a Customer to become Committed to the product or service. This happens when we pass them on to partners who can help them take the next step. This is illustrated below.

From Contact to Committed

O ──────────▶ O ──────────▶ O

Contact　　　　　**Customer**　　　　**Committed**

A STRATEGY ORGANIZED WITH PURPOSE

Certain actions must be performed to move people from one place to the other. In other words, the actions of the organization are performed in order to lead a Contact to the ultimate goal of becoming the Committed. What are these actions? They are illustrated in the diagram below.

Identifying Organizational Activity

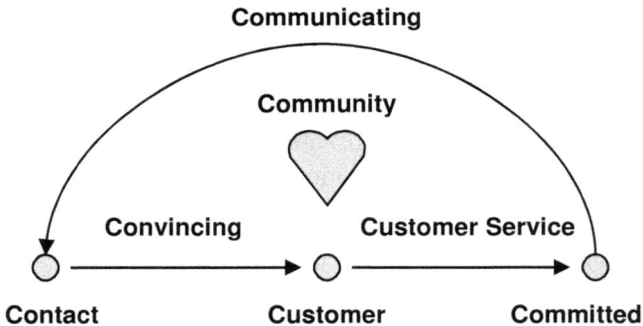

Communicating

Community

Convincing **Customer Service**

Contact   Customer   Committed

You will notice that each action is identified by a word located over the directional arrows. A "Contact" becomes a "Customer" because organizational partners "convince" those "Contacts" that the product or service which is offered is of benefit. We take action so they'll be convinced. The "Customer" becomes "Committed" to the product or service because organizational partners have provided exceptional "customer service." We take action to ensure they are served. You will also notice an arrow looping back and connecting the "Committed" back to the

"Contact." This illustrates that the Customer who is committed to the product begins "communicating" information to other Contacts whom are in need of the product or service. We take action to encourage the "Committed" to share this information.

At the center of the diagram is a heart with the word "Community" located above it. The heart of an organization is marked by the sense of community that exists between partners (workers or volunteers). This has much to do with what we learned earlier in Chapters One and Two. When all partners have the same "mindset" and "passion," it builds a sense of community. This translates into cooperation which encourages greater effort in convincing Contacts to become Customers, and better customer service that leads the Customer to become Committed. Their connection with one another and their desire to fulfill mission has a positive effect on attitude which produces the "whatever it takes" attitude we're looking for.

THE ACTIVITIES

Let's take a closer look at the actions performed by organizational partners. In looking at each, we will learn how to appropriately use them in a way that fulfills purpose and maintains a strong sense of community in the organization.

Convincing

We first notice that the Contact must be convinced. This activity is critical to the continual health of the organization and to its expansion. Many methods are used in the process of convincing a Contact to become a Customer. Business ethics come into play. The goal is not to manipulate someone into purchasing a product or service, but to show how a product or service meets a genuine need. Convincing should be "need based."

Most people have experienced what is called "buyers remorse." Many did so because incorrect information was given or pressure was felt. Someone played on their emotions. If a Contact feels manipulated, this experience will obviously lead to a bad impression of the organization. This, in turn, will keep them from becoming a satisfied Customer who is Committed enough to share information with others who are in need of the product or service. Becoming a "need based" organization overcomes this challenge. Methods are used to identify the need and to clearly communicate how what is being offered can fulfill that need in a positive way.

We have learned about needs in an earlier chapter. It is the basis for our determining the mission of the organization. We first recognize a need that can be met. The goal is to do what we can through our abilities and resources to meet that need. If the organization's mission is need based, it is natural to focus on customer's needs. If the motivation is profit, manipulation easily becomes apart of the culture of the organization. The desire is always to improve the bottom line. The goal should be to meet needs and become profitable, not become profitable at the expense of meeting needs. When we have the right approach, we change how we speak. We stop saying "I want you to buy this!" to "I want to help you."

Customer Service

The key word in Customer Service is the word "service." This comes from the word "serve." To "serve" means "to assist." Although the product or service may meet a need held by the Customer, other needs may arise. Knowing this information is critical in serving the Customer. Affective Customer Service requires continual contact in order to gain information about how the product or service is being utilized and if it is meeting their needs. If it is not, the goal is to assist them in order to meet their concerns. After all, we're a "need based" organization.

Unfortunately, many organizations only provide Customer service when problems come about. These problems are discovered when the Customer contacts the organization and shares their concern or complaint. No effort has been made by the organization to make contact with the Customer prior to this event to build the relationship and to offer assistance if needed.

Organizations that wait to be contacted are "reaction based" and those who attempt to discover issues first are "initiation based." Clearly, it is not always possible to discover problems first by contacting the Customer. However, if previous communication has taken place with the Customer by partners of the organization before problems occur, a much better relationship between a Customer and organization can be expected.

Communicating

As mentioned before, the ultimate goal is to lead a Customer to become the Committed. Those who have a good experience with the product or service become Committed and often turn into loyal Customers. The more loyal the Customer, the greater the potential that positive information will be shared with others. This information is provided through communication.

One of the best measurements for the success of an organization is through listening to what people who have purchased their product or service are saying. Word of mouth can be your best friend or worst enemy. Since this is the case, organizations should be making efforts to discover what is being said about the product or service provided.

A practical way of accomplishing this is by surveying those who have purchased your product or service by asking them how willing they would be to encourage others to purchase what you offer. The question that should always follow this is – "Why?" Understanding why people are willing to say positive or negative things about you gives you great information as we measure our effectiveness. Another important question to ask is "How did you find out about our product or service?" If they aren't talking about people who are current customers who gave them information, you might not be doing a good job in customer service.

Individuals or organizations who have personal experience with the product you are selling are your strongest communicators. Those who have responsibility to sale the product or service in the organization have a communication problem when they have not made a purchase themselves.

Several years ago, I decided to purchase a motorcycle and visited a local Suzuki dealership. They had some excellent bikes that were of interest to me. The salesperson was very helpful and informative about the products. He was also a motorcyclist, which gave him more credibility. I asked him what type of motorcycle he owned. He answered – "A Harley Davidson!" I found it very interesting that someone who was attempting to sell a Suzuki actually owned a motorcycle produced by another company. Guess what type of motorcycle I bought? A Harley Davidson!

I share this story to help us understand the importance of our buying into the product or service ourselves. Not only do we have additional expertise because of our knowledge of this product or service, but we also are enjoying its benefits and are satisfied Customers. We are apart of the community. We are people who've been personally affected by the product.

If there is any company who understand the power of ownership, it's Harley Davidson. They have created a community that includes people of many different backgrounds, socio-economic positions, and languages. All have one thing in common. They own a Harley! Most are like me. They love to tell people about it. That's the power of personal experience with a product. If we believe in it, we'll sell it.

ORGANIZING IN ORDER TO ACCOMPLISH PURPOSE

The diagram that we used to learn about the activities to move a Contact to be a Customer, and a Customer to be one of the Committed is also used to organize what we do to move people along this path. What are you doing to move people from one location to the other? It is important to have a plan.

You'll notice the diagram below called "The Organization Of Activities." This illustrates how we structure what we do to become successful.

The Organization of Activities

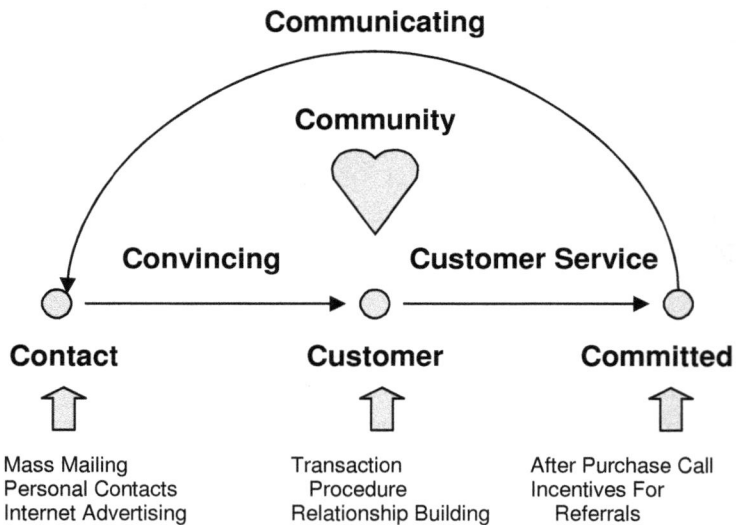

Communicating

Community

Convincing   **Customer Service**

Contact   **Customer**   **Committed**

Mass Mailing
Personal Contacts
Internet Advertising

Transaction
Procedure
Relationship Building

After Purchase Call
Incentives For
Referrals

Notice that the words "Contact," "Customer," and "Committed" have arrows pointed to them. The purpose of these arrows is to write under each what your organization is doing to connect with or serve that group of people. For example, under "Contact" you may write "mass mailing," "personal contacts," or "internet advertising" as ways in which you are reaching new contacts. Under "Customer" you may write information about the process you follow in making transactions or what you do to build relationships with new customers. Under "Committed" you may write information about what you do to contact customers after the sale or after they have secured your services. Incentives may also be offered to them when they share information with other Contacts. You get the picture! In the "Questions" section of this chapter, you will have an opportunity to write the activities you perform in a separate diagram.

Now that we have a picture of how to plan our activities, it is important for us to begin looking at what we are doing to determine whether or not our efforts are successful. It's time to be an "organizational mathematician." To do this, answer the following questions.

- Are we doing the right things?
- Are we doing the wrong things?
- Are we doing enough things?

The things we are doing that are right, we should keep. We need to think of this in two ways. It may be that we're doing the right things well. If so, we need to keep doing them the way we've been doing them. It may be that we are doing the right things but are doing them poorly. If this is the case, it's time to adjust.

Let's address the other two questions. The things that we are doing wrong, we should cut. The things that we aren't doing that would make a difference, we should add.

The success of the leader is dependent upon, not only recognizing the changes that need to be made, but the courage to make those changes. The success of the organization is also dependent upon it.

Take some time to answer the following questions to help you develop your organizational plan.

Questions

- How effective is your organization in convincing those who are in need of your product or service to commit to purchasing what you offer? Why do you feel this way?

- What weaknesses do you see in your current efforts of convincing people to purchase what you offer?

- How effective is your organization in providing quality care for those who are Customers of your organization? Why do you feel this way?

- What weaknesses do you see in providing quality care for those who are currently Customers?

- Are your Customers sharing positive information with others in the community about the product or service you provide? Why or why not?

- One of the strong selling points for Harley Davidson is the community they have created centered around their products. How could you create a community that centers around the product or service you provide?

- What specific activities are you performing to help in the success of the organization? Use the diagram below to answer this question. Under "Contacts" write what your organization is doing to reach potential Contacts. Under "Customers" write what you are doing to ensure that new Customers are having a positive experience. Under "Committed" write what you are doing to provide quality care after their commitment to purchase your product or service?

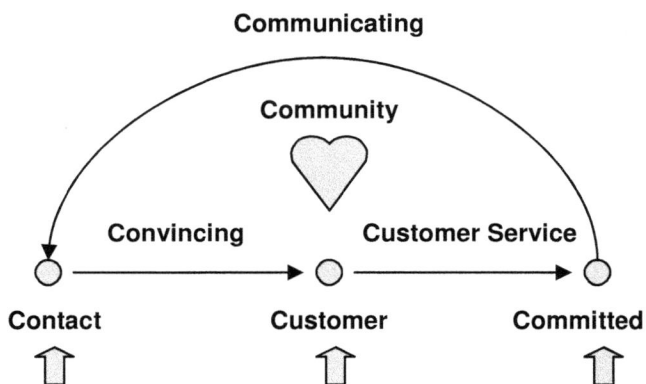

- What activities listed above need to be removed because of their ineffectiveness?

- What activities listed above need to be adjusted because of their ineffectiveness?

- What activities listed above need to be added to improve your effectiveness?

Notes

(Other thoughts related to what you have learned in this chapter)

5

The Person

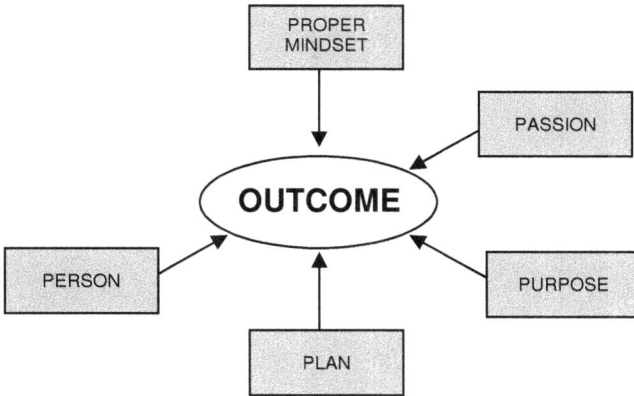

PROPER MINDSET

PASSION

OUTCOME

PERSON

PLAN

PURPOSE

---

*T*he hot Florida sun is shining brightly with no clouds in sight as you wait your turn to tee up a little white dimpled ball. You are not the greatest golfer in the world, but becoming a pastor has been quite helpful to your game. You normally play with church members, which means that you can no longer throw golf clubs, scream, or pick up your ball in anger and throw it in the lake. Now you must show patience and control your irritation. You've discovered that the golf course has a funny way of revealing a person's true character, even yours.

You wonder what this hole will bring. Success requires a long shot with your driver down the middle of the fairway. If you miss it, large sand-filled traps are to the right and the left. If you hit it too long, water from the lake comes into play.

You hit the ball, and surprisingly it stays just on the right side of the fairway, a perfect location for your next shot to the green. The decisions become more difficult now. Choosing the driver off the tee was a no-brainer. You love the driver. Now you must decide which iron to use. This is difficult because you never seem to hit your irons the same way twice.

You are about 100 yards from the green and know that if you hit your 9 iron smoothly, you'll be right on the hole. The problem is you only hit your irons smoothly about 20% of the time. You must decide whether to compensate for past failures and choose an 8 iron instead, hoping that you don't hit it perfectly. If you do, you'll hit the ball 10 yards past the green and into the woods. You decide to play your 9 iron and hope for the best. You swing, and to your great astonishment, you are on the green just 15 feet from the hole. The fun begins!

The Person

You now have the great privilege of using your putter, the dreaded putter. The good news is that you haven't purposely broken one since you became a pastor. The bad news is that you previously disposed of three. The pressure is on. You putt...

The Leadership Diamond

The purpose of this chapter is to begin developing in leaders and organizational partners qualities that are necessary for success. Understanding your mission and having a plan is not enough. Poor leadership causes instability in an organization. John Maxwell, a noted author and speaker in the area of leadership, wrote: "No matter how hard you work, you can only go so far professionally if you are a poor leader. The company, department, or team will always be held back by a weak leader."[1] This applies to all organizations.

Becoming an effective leader isn't about just one thing, it's about several. To help you grasp this, we'll be looking at issues that affect who we are as leaders. You'll notice that they include specific areas. There are four components that affect the leader's ability to influence others to accomplish its stated mission. These are illustrated in the following diagram.

LEADERSHIP
STYLES

LEADERSHIP   PERSONAL
TRAITS   STYLES

FUNDAMENTAL
CAPABILITIES

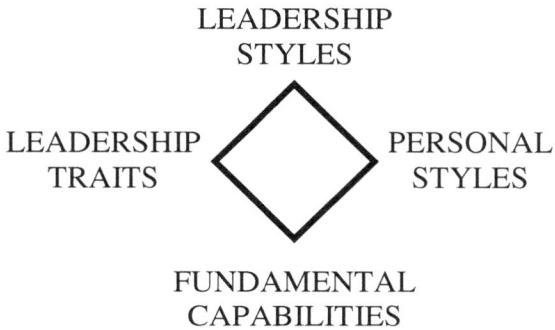

THE SKILL OF CHOOSING LEADERSHIP STYLES

Choosing which golf club to use on the golf course is similar to choosing a leadership style. The situation dictates which is needed. Daniel Goleman, a professor at Harvard, taught about the importance of choosing the appropriate leadership style for a given situation in an article in the *Harvard Business Review.* He wrote:

> ... the research indicates that leaders with the best results do not rely on only one leadership style; they use most of them in a given week - seamlessly and in different measure - depending on the business situation. Imagine the styles, then, as the array of clubs in a golf pros bag. Over the course of the game, the pro-picks and chooses clubs based on the demands of the shot.[2]

Are you a pro, or are you like the golfer who has a difficult time deciding which club to use? Consider how ridiculous it would be for a golfer to use his driver on every shot, whether he was far from the green, close to the green, or on the green. Yet some leaders use the same leadership style in every situation, even when it is inappropriate. Healthy organizations have leaders who understand their predominant leadership styles and are adept at using the appropriate style when necessary. What are the styles?

LEADERSHIP STYLES

Goleman lists six leadership styles, defines them, and deals with situations that are appropriate for their use. Take a look at what they include. [3]

The Visionary Leader

He leads people toward a vision. The style works best when an organization needs a new vision or when a clear direction is needed.

Former Whitehouse official Charles Colson wrote about the power of vision in his book *How Now Shall We Live.* He told the story of a man who brought positive change into a difficult community. He wrote," On his way to work one day, Chicago insurance broker Bob Muzikowski saw a derelict ball field full of trash in a gang-infested neighborhood. *The kids there could use a real Little League to play in*, he thought. He teamed up with a friend to create the Near North Little League. In "pretty wild" early practice sessions, coaches dealt with 250 boys long on enthusiasm but short on fundamentals." Because of his belief that he was meant to serve and make a difference, he committed himself to invest his life in these young men (who at times seemed to be out of control) to help impact their future. It worked! Colson pointed out that 400 kids joined the league the next year and that today 900 are involved in 100 teams. All of this happened because of the vision of one man. Vision can make a difference![4]

The vision of this man became his mission. The word vision and mission are closely related. We've learned about mission earlier in the book and discovered that it is the catalyst for what we do. The visionary leader is the one who is sensitive to the needs that are in the community that need to be met. This sensitivity leads him or her to begin dreaming the dream of what could be done to make a difference. The dream becomes a mission that must be met.

The Coercive Leader

He demands immediate compliance from others. The style works best when in a crisis, to kick-start a turnaround, or with problem employees.

When I think of this type of leader, I think of a drill sergeant in the military who trains his troops. Although they may not want a friend who is so forceful in mannerism, actions and words – there are times that they need someone to exhibit this type of behavior. This leader instills what is needed to make it through the difficulties that they could one day face. They prepare them for a crisis. In organizations, there are times that leaders must be firm in their decisions and also in the way they react to organizational partners to ensure success in stressful situations.

The Affiliative Leader

He creates harmony and builds emotional bonds. The style works best to heal rifts in a team or to motivate people during stressful circumstances.

In religious circles, we would call this a "kum-ba-ya" leader. This is someone who likes to sit around a campfire, tell nice stories, and sing songs. They want everyone to get along. If conflict occurs between partners, they do everything they can to bring unity once again. They are the peacemakers. There is a place for this. We need to be concerned with how we interact with one another and, as a leader, we must deal with situations when relationships are strained. We all need our "kum-ba-ya" moments.

The Democratic Leader

He forges consensus through participation. The style works best to build buy-in or consensus, or to get input from valuable employees.

The word democratic means "social equality for all." This leader values the opinion of others and wants everyone to have a voice. They recognize the power that comes when new ideas are shared. They also know how the involvement of partners affects their buy-in as movement is made toward accomplishing organizational plans. Although being a democratic leader can be very useful, there are times that consensus does not occur. In this case, the leader has two options – wait until there is agreement or make a decision that best fits the mission and purpose. Unfortunately, some organizations wait for consensus and miss a great opportunity to act. The reaction to this missed opportunity is regret and possible anger at the leader who would not courageously make a difficult decision.

Often, this leader fears failure. They think – *if we make a decision as a team, then we are to blame. If I make a decision as the leader, then I am to blame.* The leader cannot be afraid to fail and be blamed. After all, one of the greatest education opportunities is learning from the mistakes that we make. It brings maturity and growth.

Those who grow the most are willing to take the risk. This reminds me of what Michael Jordan shared about his own failures. He said, "I have missed more than 9,000 shots in my career. I have lost almost 300 games. On 26 occasions I have been entrusted to take the game winning shot and I missed. I have failed over and over and over again in my life. And that's precisely why I succeed."[5] As Jordan has learned, our greatest failures can lead to our greatest successes.

The Pacesetting Leader

He is driven to achieve and sets high standards for performance. The style works best to get quick results from a highly motivated and competent team.

This leader has a need for speed much like a NASCAR vehicle. A Physics professor at the University of Texas named Diandra Leslie-Pelecky reflected on a NASCAR racer's need for speed in her book *The Physics of NASCAR.* She was given the opportunity to drive a NASCAR car on the 1.5 mile track of the Texas Motor Speedway with an instructor by her side. Her top speed – 150 miles per hour! It was the process of accelerating to that speed that she learned something interesting about race cars. She writes, *"We trundled down pit road, and when Paul motioned, I pressed the clutch, shifted into third, then released the clutch and stepped on the gas... A NASCAR engine is optimized for speed, so when you're puttering along at 100 mile per hour, it chugs uncomfortably. The solution is to go faster.*"[6]

There are some NASCAR leaders out there who thrive on doing things at 150 miles per hour. They like going faster. Pacesetting leaders are highly motivated people who do things more quickly than others. They typically work longer hours and are high producers. There are times that this is very useful. If you are working on a timeline and need quick results, you need this type of leader. The challenge for those who deal with this type of leader is that they can't keep up and may feel like they've been left behind. This often causes them to feel threatened because of their belief that they can't measure up. It can lead to a bad attitude toward a leader and division between partners.

The Coaching Leader

He develops people for the future by investing time in them. The style works best to help an employee improve performance or develop long-term strengths.

One of the greatest coaches in basketball history is a man named John Wooden who coached for UCLA. He won ten NCAA basketball championships in 12 years. One of the most popular players who had the privilege of being coached by him is Bill Walton. He played for Wooden during a stressful time in our country's history. The war in Vietnam was all over the news and Nixon was being challenged because of his role in the Watergate scandal. For Wooden, the answers never changed. Walton said, "We thought he was nuts. But in all his preachings and teachings, everything he told us turned out to be true... His interest and goal was to make you the best basketball player but first to make you the best person. He would never talk wins and losses but what we needed to succeed in life. Once you were a good human being, you had a chance to be a good player. He never deviated from that. He never tried to be your friend. He was your teacher, your coach. He handled us with extreme patience... He didn't teach basketball. He taught life..." He went on to say, "When you're touched by someone that special, it changes your life. You spend your life chasing it down, trying to recreate it."[7]

A great coach, like Wooden, is first concerned about developing us into the best person we can be. Our ability to be successful is directly connected to who we are as people. Coaches have a vision for our potential and have the goal of helping us achieve it. They are able to be patient when patience is required and firm when discipline is needed. Coaches aren't people who want to hold on to their position's, they are people who want to develop others to the point where they can take over. The heart of a coach is one of sacrifice and service.

You have most likely been able to identify your predominant leadership style by reading these definitions. Do you use this style most of the time, or do you change styles according to the situation? Each style has its place in an organization. The more professional you become at using them, the more successful you will be. The health of the organization is dependent upon wise choices by you, the leader.

PERSONAL STYLES

I'm a big American Idol fan. This might stem from my unfulfilled dream of being a recording artist and big star. I've shared with my kids many times that if I would have only been born 10 years later (maybe 15) I could have been in N'SYNC. As you can tell, I may have an elevated view of myself.

Getting back to American Idol! The judges have changed over the past three years from the original cast. You may or may not like those changes. I, for one, really dig the cast of today – especially Steven Tyler. I have my reasons. I think it's the hair! The cast of judges that most are familiar with are the original three. It included Simon Cowell, Paula Abdul and Randy Jackson. All have very different personalities.

In an article by Paul Coughlin, he emphasized the differences. He used them as examples to teach the importance of avoiding passive and aggressive extremes and to choose, instead, assertiveness. He wrote, "Three major personality types are found among the judges of the popular reality TV show *American Idol.* Passive Paula Abdul is gracious but not always truthful. Aggressive Simon Cowell is truthful but rarely gracious. Assertive Randy Jackson is often truthful and gracious. Be like Randy!"[8]

Okay, we know Randy isn't perfect either. However, what Coughlin wrote has some validity. We need to be truthful but also gracious. In other words, we need to tell people what they need to hear in the way they need to hear it. Leaders aren't to crush spirits, they are to help those who follow see reality and gain hope. This hope comes when we recognize our strengths and the limitations of our weaknesses. Hope comes when we match our dreams with the reality of who we are. Good leaders help others recognize how fortunate they are to have the strengths they have and to also speak truth about areas of weakness. There are times that we can overcome them, yet there are other times when much energy is placed in areas that will not change. So what should we do? We need to focus on our strengths and continue to develop them.

What are your strengths related to your personality? Do you know what they are? Fortunately, there are tools that help us determine this. One method has become very popular in organizational circles. It's called the DiSC profile which was developed by Carlson Learning Systems.[9] The acrostic identifies four specific personal styles that are defined below. As you read through each, circle the words or statements that best describe you.

Dominance

- Risk takers
- Determined
- Decision-maker
- Competitive
- Problem solver
- Productive
- Enjoys challenges
- Goal driven

Interaction

- Energetic
- Very verbal
- Spontaneous

- Friendly
- Optimistic
- Thinks out loud
- Popular
- Motivates others

Steadiness

- Loyal
- Avoid confrontation
- Dislikes change
- Patient
- Sympathetic
- Indecisive
- Sensitive
- Not demanding of others

Cautiousness

- Accurate
- Practical
- Reserved
- Orderly
- Factual
- Likes instructions
- Detailed
- Conscientious

Which of these personal styles best describes you? Understanding your personal style is not enough to experience success as a leader. We must also master four fundamental capabilities.

THE FOUR FUNDAMENTAL CAPABILITIES

Understanding predominant leadership styles and personal traits is important because they impact the way we interact with others. Goleman taught that leaders must also consider our "emotional intelligence." He wrote that emotional intelligence is "the ability to manage ourselves and our relationships effectively."[10] Our doing this is dependent upon what he calls the "fundamental capabilities." These capabilities better describe what makes up our emotional intelligence. They include:

Self-Awareness

The self-aware have the ability to read and understand their emotions as well as recognize their impact on work performance, relationships, and the like.[11]

Those who are self-aware know their strengths and weaknesses. They know what they can do well and what are challenges. Our ability to grow as leaders is dependent upon our self-awareness. If we don't believe there is room for improvement, there is nothing to grow. All of us have room for improvement and need to be willing to take a hard look at self to discover where the improvement needs to take place. There are times that we don't see things about ourselves that are detrimental to our success. Knowing that this can be the case, it is important for leaders to partner with others who are willing to share information with us that will help us. It begins with their willingness to point out areas of weakness that we might not see. I am grateful to have a wife who openly shares information about me that I don't see.

Years ago, Jennifer had one of these opportunities. I have a tendency on the weekends to be a very tunnel vision person. My mind is on speaking and I'm concentrating on what I am going to say. My habit was to leave my office and walk directly to the meeting area where I was to speak, making eye contact with no one. I didn't do this deliberately, my mind was simply on what I had prepared to share with those who attended our services. In the process of that walk one day, I passed directly by someone who called my name and said hello. Jennifer witnessed the event.

When we arrived at home that afternoon, she told me what I had done. She also told me that I had better stop ignoring people. In a way that I would receive it well, she explained how people felt when I didn't acknowledge them. She was just as concerned about how I would respond to what she was telling me as she was for the one I had offended. Because of her, I became self-aware and changed my behavior.

Self-Management

Those who practice self-management have the ability to keep disruptive emotions and impulses under control.[12]

People don't respect people who are out of control. This is a trust issue. Partners in the organization want to know that we won't fly off the handle or do something irresponsible because of a stressful situation that may arise. It may be that we would have done so at one time but became self-aware about our responses and decided to change. The natural follow-up to self-awareness is to learn how to manage ourselves in areas of our weakness. This is growth.

Social Awareness

The socially aware are skilled at sensing other people's emotions, understanding their perspective, and taking an active interest in their concerns.[13]

In social awareness, we must be people who pick up on social cues. We need to be able to sense when we are making people uncomfortable or are irritating them. There are those who have no clue about the annoyance people feel. The primary reason for this is that the annoyer isn't thinking about the other person. They may be looking in their direction, but they really aren't noticing them. They are too busy thinking about what they are going to say or do next. People don't socially connect with those who only talk about themselves and their personal interests. Partners need to know that we care about them. They cannot if we don't connect with them on a personal level. That connection requires engaged two way conversation.

Social Skill

Those who use proper social skill have the ability to take charge and inspire with a compelling vision through the use of good communication.[14]

Social Skill is the natural follow-up to Social Awareness. If we take time to notice others and really listen to them, we begin to understand their perspective. It's only when we understand where they are coming from that we have the information that we need to address their concerns. These concerns may be keeping them from buying into a vision that we are casting for the organization. It won't be compelling until their concerns are no longer an obstacle. The person who has social skill addressed them in a way that obstacles are removed.

As you look back on the definitions of the Four Fundamental Capabilities, how would you score yourself related to your abilities? Using a scale of 0 to 10, with 0 meaning that you are completely unsuccessful and 10 meaning that you are completely successful, rate yourself next to each of the capabilities listed below.

- Self-Awareness = _____
- Self-Management = _____
- Social Awareness = _____
- Social Skill = _____

The final component that affects our ability to lead is called...

LEADERSHIP TRAITS

What do we look for in a leader? Through my personal experience and study I have discovered certain traits which impact our influence. Below is a listing of those traits and a description of each.

They are capable.

The word capable means "having the ability or qualities necessary for" Those selected to lead are to have the abilities and qualities necessary to accomplish the assigned tasks. Do you have the abilities to do your assignment?

They are trustworthy.

The word trustworthy means "dependable and reliable." Those selected to lead are to be dependable, having the trust of others. Can you be trusted to follow through on your commitments?

They are an example in life.

They provide the proper example by illustrating to others proper work ethic and how to communicate with others in a positive way. Are you an example in your work and through your words?

They teach others how to perform their tasks.

They are to properly communicate responsibilities and provide instruction so that organizational partners can achieve those tasks. Do you train others to do the jobs they have been assigned?

They select other leaders.

They choose people on the basis of the following:

- *Character*. They care about others.
- *Competency*. They are capable to perform their tasks.
- *Commitment*. They believe in the vision.
- *Chemistry*. They cooperate with others and are helpful to the team.

Do you recruit other leaders who meet this criteria?

They resolve conflict.

They are able to handle difficult situations that those in lower levels of leadership cannot. They are observant - noticing areas of weakness, and deal with those situations in a timely manner. Do you resolve conflict before it becomes an emergency?

They delegate.

They do not show that they are protective of their position by holding on to tasks that others can perform. They encourage others to improve on their abilities by giving them new tasks. They also allow them to do the jobs they have been assigned. Do you allow people to do the jobs they have been assigned?

Do all of these traits describe you? A healthy organization has leaders who are defined by each. One way to determine if these traits are true of you is to restate them in a personal manner. Complete the following exercise.

LEADERSHIP EXERCISE

Using a scale of 0 to 10, with 0 meaning that you are completely unsuccessful and 10 meaning that you are completely successful, rate yourself according to the statements.

I am capable. _____

I am dependable. _____

I am an example for others in life. _____

I teach others how to perform their tasks. _____

I select capable people to assist me. _____

I resolve conflict. _____

I delegate responsibilities and allow those under me to do their jobs. _____

WHERE'S THE BALL

We ended the opening illustration with a putt. Is it in the hole? Mastering the leadership traits makes you a pro and puts the ball in the center of the cup. If you missed, don't get frustrated with the game and give up. Keep practicing and learn to be the best leader that you can be. Answer the following questions to better understand your personal leadership strengths and weaknesses.

Questions

• Has there been a time when you have been a Visionary Leader? If so, describe it.

- Has there been a time when you have been a Coercive Leader? If so, describe it.

- Has there been a time when you have been an Affiliative Leader? If so, describe it.

- Has there been a time when you have been a Democratic Leader? If so, describe it.

- Has there been a time when you have been a Pacesetting Leader? If so, describe it.

- Has there been a time when you were a Coaching Leader? If so, describe it.

- Which of the personal styles - dominance, interaction, steadiness, cautiousness - best describes you? Why do you believe this is true?

- Of the four fundamental capabilities (self-awareness, self-management, social awareness, social skill), which do you need to work on to improve? Why?

- Of the seven leadership traits mentioned, which needs the most improvement and what will you do to work on this improvement?

Notes

(Other thoughts related to what you have learned in this chapter)

6

The Partnership

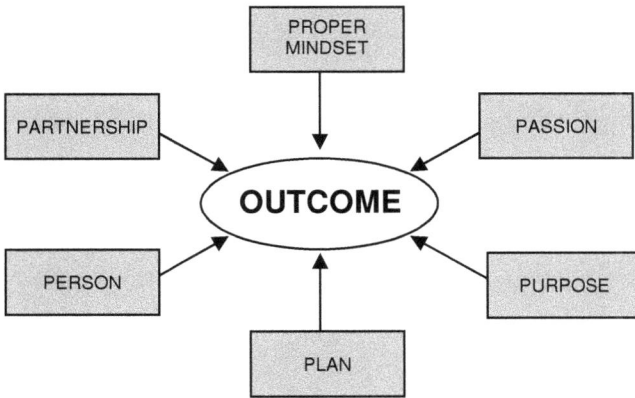

---

*I*n an article in the New York Times, David Waldstein writes about a man who made an impact for a professional baseball team that truly made a difference. He writes…

> In the realm of sports superstars, Mike Murphy plays a small but essential role for his pro baseball team, the San Francisco Giants. Recently, he helped the Giants win their first World Series in 52 years, and Mike was on the job for the entire 52-year journey. But you won't find his name on the scoreboard. You won't see him endorsing new products for millions of dollars per advertisement. He'll never make it into the Baseball Hall of Fame. "Murph," as they call him, has been with the team since 1958, starting as the team's batboy, before moving up to his roles as the clubhouse attendant and finally the equipment manager. Throughout his 52 years of faithful service, Murph has worked behind the scenes, cleaning shoes and ordering bats, so his much more famous teammates could achieve success. If anyone embodies the 52-year endurance test the San Francisco Giants underwent to reach baseball's summit, it is Mike Murphy. That's one reason why the owner of the team handed the World Series trophy to Murphy so he could present it to the ecstatic teammates.[1]

Why would the owner do this? It's simple! On a team, every job is important as we work together to win the prize whatever that prize may be.

THE FIVE QUALITIES OF A TEAM

Mike knew his role on the team and did his job well. Problems come when we drop the ball. Not only must we do what we are expected to do in our tasks, but to truly be a team, we must work together. How can we overcome the challenge of working with a group of people who don't function as a team?

Our ability to do this effectively first begins with our understanding what can happen when team chemistry occurs. The more we recognize the potential of what could be, the more motivated we are to put forth the energy necessary to help change the team dynamic.

The great benefit of developing a sense of team is the results. When two or more people work together as a team, they produce something greater than they could alone. Our results are multiplied. A common term for this is the word "synergy." This is an important term to consider because synergy is the difference between a team and a group of people. The moment synergy occurs we can legitimately describe ourselves as a team. Until that time, we are merely a group of people made up of individuals who do their own thing.

What causes us to experience this synergy? It's not just our working together, it's our level of cooperation. We may do things together with low cooperation which produce few results. In fact, there may be little difference at all. The higher the level of cooperation, the more the team energy and the greater the results.

Cooperation grows when we believe we need each other. When this occurs, we are more willing to work with those on the team. The key is working together. Although this may sound easy, it is not. We must learn how to be true partners and function as a team. So what's involved?

Quality One:
A Team Functions as a Crew

I think about a specific event that took place in my life several years ago when I think about a crew. One of my great passions is sailing. As I mentioned earlier, I have owned many boats. Three of them have been sailboats. I'm a wind guy. There is just something invigorating to me about being out in nature and feeling the breeze.

My first sailing experience occurred when I was in college. Our fraternity, Sigma Chi, had traveled to Fort Walton Beach, Florida for our annual formal. The girlfriend of one of my fraternity brothers lived in the area and her parents owned a beautiful boat. They invited us out for the day and several of us went along for what proved to be an amazing experience on the water.

Jennifer and I were dating at the time and were talking about getting married. I told her that I had a new dream. It was to purchase a boat as soon as we could after our big day. That day finally came. After our wedding we were living in Pensacola, an incredible location for sailing enthusiasts. Being there and often seeing the beautiful sails moving slowly across the water as we traveled across the bay bridge only served to stoke my passion.

I came up with a plan that included Jennifer's brother. We couldn't afford a boat on our own, but if we partnered with him this dream could come true. The sales job began and he bought into the idea. Before we knew it, we were at the marina looking for a boat that would be our own.

The vessel we decided to purchase was a twenty five foot Hunter. It was an awesome ride! After purchasing it, we had a 30 minute lesson from the sales person on the basics of sailing. The crash course ended and he sent us on our way. This was, to say the least, an interesting first day for our crew. We nearly ran into the Pensacola Bay Bridge. This was not a good omen for the future.

As time went by, our sailing abilities drastically improved. Because of our difficult beginnings, we bought a book on sailing to learn the right terminology and better understand sailing tactics. This, combined with trial and error on the boat, helped to build our comfort level.

A few weeks later we were coming in from our sail on the bay and I decided that my skills had improved enough to sail the boat up the channel which led to our dock. Big mistake! The wind was coming across the port side of the boat (just throwing out some boating language to try and impress you) and began to push us sideways toward the sandbar that lined the boating lane. Before we knew it, we were aground.

The story doesn't end there. I jumped out of the boat into freezing water (it was during the early spring) to attempt to push the boat off of the sandbar with no luck. Another boat saw our predicament and came to attempt a rescue. We threw him our line so that he could pull the boat off the bar, with me still in the water. What happened next was completely unexpected. The line wrapped around the keel of the boat, so when it came off of the sandbar, it was traveling backwards. There was no way to control the direction of the boat as it became free.

Then we heard it. There was the loud sound of the horn of a barge named "Kristy" that was now coming down the channel with no way to change course. The sailboat, with Jennifer and David aboard, was being pulled directly in front of the barge by this small boat. It looked as if a tragedy was about to happen. Fortunately, at the last moment, the small boat revved its engine and pulled them out of the barge's path with about 10 feet to spare. It was one of the scariest moments any of us have ever experienced.

Why do I share this story? It was the inexperience of the crew combined with my arrogance that almost led to disaster. If the right person were the captain of the ship he would have never attempted to sail into a channel in such difficult wind conditions. The crew really does make the difference.

Theresa Kline wrote about teams in her book entitled *Remaking Teams*. She wrote:

> Unfortunately, there is as yet no common way to talk about different types of teams. This situation, however, is changing. In the latter part of the 1990s there has been a concerted effort on the part of several researchers to tackle the task of building a taxonomy of teams. Their work suggests that there are fundamental differences between team types and that it is important to attend to them. The line of research has revealed several common threads in what differentiates team types, including the degree of structure of the task, prescriptiveness of the roles of the partners,

nature of the information exchange between partners, and type and degree of sharing a common goal among the partners. Depending on where the team falls on these dimensions, there are ramifications for where effort needs to be placed to make the team more effective.[2]

She introduced a specific type of team called a "crew." This type of team is well-defined and moves in a unified direction. It is important for us to describe what this crew looks like. We need to answer the question, "Who makes up the crew?" The answer to that question is discovered by recalling what we have learned throughout the course of this book. Remember, there are foundational posts that cause an organization to be healthy and effective. We also need to remember that the organization is made up of people. The five foundational posts described in the five previous chapters of this book are the components that must be mastered by organizational partners if they are to function as a crew. Let's summarize what must be done through some specific statements.

Each crew member must have the proper mindset.

This means that each crew member must understand that we are in this together, believe that balance matters, believe that the people we are reaching matter, and begin with the end in mind. Everyone having the same mindset brings unity and causes us to work together as we move toward the destination.

***Each crew member must be passionate
about the organization.***

They are passionate because they desire to be a movement
and not an institution. They understand that a movement
occurs when each partner lives for "we" and not "me."
They are passionate about what the organization is
attempting to accomplish to the point that they are willing
to change.

***Each crew member must know their purpose and have a
desire to achieve the ultimate destination.***

The crew members must be aligned toward fulfilling the
mission while also working to accomplish their specific
purpose. Each member must have a desire to achieve it.
MacMillan, while teaching on teamwork, used the
illustration of a boat with crew members rowing in the
same direction. Check out the diagram.[3]

The Crew Aligned With Purpose

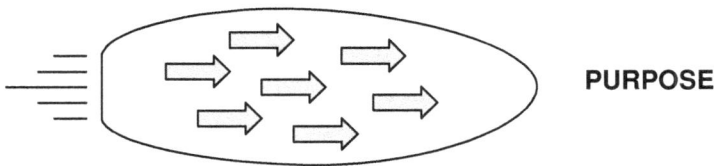

PURPOSE

The diagram above reveals that the crew is working
together toward a common purpose. However, problems
arise on the team when its partners begin rowing away from

the team's purpose. The following diagram shows that the crew is in trouble.

The Crew Misaligned With Purpose

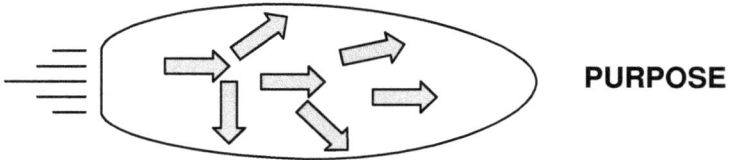

Three crew partners are rowing toward the team purpose, and four crew partners are rowing in alternate directions. Which crew member causes the team the most difficulty? Some may think it is the one rowing in a direction farthest away from the team purpose. Actually, this is the easiest problem to correct. His direction away from the team purpose is so obvious that the other crew partners can just kick him off the boat (please don't forget the life jacket). The crew partners who are rowing slightly off course can cause the most difficulty because they are hard to detect. They bring just enough friction to cause the team to miss its destination. Why does this happen? What causes us to get off course? Often, we stop believing that we need each other.

In his book *Outliers,* Malcome Gladwell wrote about an amazing man named Christopher Langan. What made him unique was his incredible intelligence. He was a true genius with an IQ of 195. To help you understand just how impressive this is, Albert Einstein's IQ was 150. Langan could learn and retain information quickly. An example of this was his ability to ace foreign language tests by skimming the textbook 2 – 3 minutes before the exam.

Another great feat was his acing the SAT even though at one point he fell asleep.

You would think that someone like this would have made great discoveries or would have risen to the top of industry, making a significant impact in the business world. Things turned out much differently for him. Instead of progressing and using his intellect to make great advancements, he ended up working on a horse farm in rural Missouri. Gladwell pointed out the problem. Langan never had a community of people around him to help him capitalize on his abilities. He lived his life alone rather than connecting himself with others. Gladwell made a great summary of Langan's life writing that Langan "had to make his way alone, and no one – not rock stars, not professional athletes, no software billionaires, and not even geniuses – ever makes it alone."[4]

Often, organizational partners don't partner together because we don't see the need to connect with others in this way. They go it alone. Why? It may be their arrogance, believing that they don't need anyone else. If we are working to achieve something that we don't believe we need help accomplishing, what we are attempting to do is not big enough. We should desire to do things that are greater than what we can do alone. The mission should be so great that it requires our willingness to work with one another. It should also be so big that we want to help each other become all they can be so that we can be successful together.

*Each crew member must recognize
where they fit in the plan.*

Not only must they know where they fit in, but also must clearly understand their specific responsibilities. It is very difficult to do well in accomplishing our purpose (the goal that we are to accomplish in our specific area) if we don't know what we are expected to do to see that it happens. Not only must we know where we fit in on the team, we must also feel like we are a significant part of the team.

*Each crew member must have the ability
to fulfill their responsibilities.*

They may know the organizations mission and buy into it and know their specific purpose and how to accomplish what is expected and still fail. The reason for this failure is their inability to accomplish the task. Do you remember the Leadership Diamond and the four components that make up who we are as a leader? They include leadership styles, personal styles, fundamental capabilities, and leadership traits. Because we are unable to accomplish our tasks, there is a weakness found in one or more of these areas. To be an effective team, all members must have the right capabilities. This is illustrated by looking at the boat once again.

The Crew Challenged By Lack of Abilities

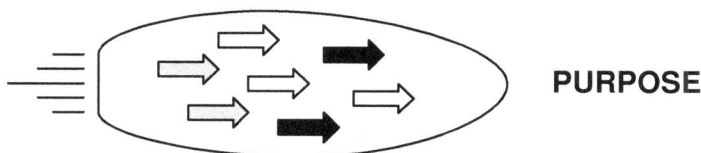

PURPOSE

The "white" arrows represent people who are very capable. They are great at their tasks and exhibit excellent leadership qualities. The "gray" arrows represent people who are partially capable to fulfill their tasks. They are partners who do some things well but other things poorly. The "black" arrows represent people who are incapable of doing what has been asked of them. Just as it is easy to decide who to kick off the boat in reference to those who are moving away from purpose, it is also easy to decide who to kick off of the boat when people are incapable of doing their work. The choice is whether or not to completely cut ties between them and the organization or to find a place within the organization that matches their abilities. Removing them from the team where they are mismatched is critical for success.

The organization's success is also dependent upon what you do with the "gray" partners. To move in a more productive way toward fulfilling purpose, there must be improvement. This requires the leaders who are responsible for fulfilling purpose to invest in them to help bring growth. If this is not done, not only will the organization be held back as it works toward accomplishing mission, it also runs the risk of those who are "gray" developing negative attitudes. Personal growth helps to prevent this. When there is growth, there is hope for greater success personally which leads to a more positive experience for the partner.

Not only are we to function as a team working as a crew, we also see another important aspect that leads to our effectiveness in working together.

Quality Two:
A Team Communicates Well

A major challenge that causes stress within the team is difficulty in communication. The area of communication is one of the major factors in the downfall of relationships in general, whether people are friends, coworkers, or are marriage partners. It is important to master some specific areas of communication if we are going to work well as a team. Let's take a look at the communication elements that make a difference.

Each team member must understand how communication works.

Often, relationships are strained because a message is not shared in an understandable way. The message may also not be received well because of the words used or the tone in which it is given. To better understand this, we need to learn about what occurs when we share information with one another. Let's take a look at the process that occurs when we communicate.

The Communication Process

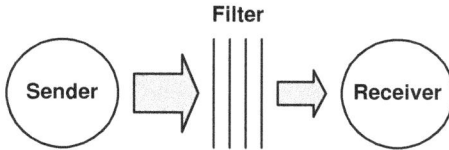

The illustration reveals two individuals: a Sender and a Receiver. The message is represented by arrows. Notice that the arrow coming from the "Sender" is larger than the arrow coming toward the "Receiver" on the right side of the filter. The message changed. Why? The Receiver filters information through his past experiences, his moral compass, and his impression of the Sender. These are the basic areas that need to be addressed if communication is to improve.

The Receiver's past may include a bad experience with an organization or a leader. He or she receives information through this filter because of this. Effective leaders must have a good understanding of the Receivers past if they are going to deliver a message in a way that is meant to be heard.

The Receiver may also have a different view of life. If they are living for what they gain and not what they give then information is filtered through this view. The Sender must know what motivates in order to communicate a message in a way that is well received.

The Receiver may also have a view of the Sender that is negative. This leads him or her to have an attitude toward the message simply because of his or her belief about the one giving the information. Once again, the Sender must be aware of this and communicate the message in a way that overcomes this attitude.

These challenges in communication may be a key factor in causing people to veer off course away from team purpose. The more partners misunderstand the purpose because of poor communication, the more likely they are to begin developing negative attitudes. These attitudes affect a team's ability to succeed.

Each team member must recognize the power of body language.

My undergraduate degree is in psychology. I remember during one particular class, we spent some time learning about how our mannerisms affect the message that we are attempting to communicate with others. I came to know this first hand just a few days after learning about how our body language can make a difference.

Jennifer and I were in college at Samford University in Birmingham, Alabama (Go Dogs!) and were excited to have a break. We decided to make the long journey from Birmingham to her parent's home just north of Tampa in a place called Spring Hill. On our way, we traveled down Highway 19 which runs north and south down the west coast of Florida. We had just passed through a small town called Cross City. After making it through the city limits, the speed limit increased and I began to speed up. In the midst of doing this, Jennifer and I were in a deep conversation. I wasn't paying attention to how fast I was going. Not only had I reached the speed limit, but I had exceeded it by about 15 miles per hour. I was going fast enough to pass another car that was also speeding.

Before I knew it, blue lights were behind me and the siren was sounding. I pulled over and the officer came up beside my car in his, which I thought was unusual. He rolled down his window and told me to stay where I was. Immediately, he took off after the car I had just passed. We could see him down the highway with the other driver, obviously giving a ticket.

While I was waiting for the officer to return, I remembered what I had just learned in my psychology class about the power of body language. So I decided to try it out. When the officer returned to deal with my violation, the experiment began. I showed him the palms of my hands over and over again. I had learned that showing your palms was a sign of honesty. The conversation took place and I shared my story about not paying attention. I also played the "couple" card saying we were on our way from college to my fiancé's parent's house in Spring Hill (with palms continuing to be pointed toward him). He looked at me, told me to stop speeding, and got back in the car. No ticket! My strategy worked. Don't miss this! He gave a ticket to the driver I had passed and let me off with a warning. It happened all because of the power of body language.

I'm not sharing this with you to help you get out of trouble when you're dealing with a police officer, I share this to say that our body language matters. Our hand motions, facial expressions, and even how we position our bodies when we speak to others, all communicate a message. They either match the words of our mouth or they say something contrary to the words we use.

Each team member must be
willing to be honest.

Many are not honest because of the fear of conflict. Rather than dealing with challenging issues when they occur, they allow them to fester. It isn't until they become emergencies and force their intervention that they get involved. Conflict is an important part of communication that leads to our success. Our dealing with conflict is often connected to our courage. This is a true test of leadership.

Please don't read into the quote I'm about to share as any affiliation with a political party, but I l love what John McCain shared on the topic. He said, "Courage is like a muscle. The more we exercise it, the stronger it gets. I sometimes worry that our collective courage is growing weaker from disuse. We don't demand it from our leaders, and our leaders don't demand it from us. The courage deficit is both our problem and our fault. As a result, too many leaders in the public and private sectors lack the courage necessary to honor their obligations to others and to uphold the essential values of leadership."[5]

If the leader is unwilling to confront situations that need to be confronted because of a fear of conflict, it reveals a weakness that decreases confidence from those who are expected to follow. However, if a leader acts with courage and is willing to face conflict head on, it encourages confidence in others to share openly their concerns. This leads to the resolution of challenges that have led to ineffectiveness. It allows the organization to move in a more positive direction.

Quality Three:
Team Members Trust One Another

You may not have heard of the pro football player named Tony Richardson, but his impact as a teammate has been impressive. His primary role doesn't involve running with the ball, but blocking to help those who do become successful. There is a great bond that has taken place between him and those he helps because of this. The players who rely on him have developed a great sense of appreciation. They also place a great deal of trust in him because of his ability to clear the way and his willingness to give of himself for their benefit.

Over his seventeen year career, he has played with some of the best players in the league. The year 2001 was to be big for him as he was expected to be the main running back. Things didn't turn out that way. He didn't lose the position because of a lack of ability, instead, he went to his teammate Priest Holms and told him "It's time for me to step out of the way. You need to be getting the ball. And I'm going to do everything I can to help you." This sacrificial decision paved the way for Holmes to lead the league in rushing. Tony didn't become bitter or jealous over his success, rather he celebrated his accomplishments. Holmes told others how Tony would call him up and say, "I just saw you on *SportsCenter!*" He said that Tony was happier for me than I was for myself.

All of those he helped succeed say that his influence went beyond blocking for them. They share how he would constantly talk to them throughout the game, giving them advice and pushing them through encouragement and inspiration. In an interview, Tony said, "I can't explain it, but it just means more to me to help someone else achieve glory. There's something about it that feels right to me."[6]

Can you imagine working with people like him? What if all of us had the same desire for those we work with to succeed. The result, just like in Tony's case, would be trust. The breakdown of trust often happens because we are unwilling to make sacrifices for the benefit of others.

Trust is the foundation on which teams are built. If trust does not exist, there will be no success. Team partners must know that each person cares about everyone on the team. This comes from their treatment of one another. If they become self-centered, they begin to act out in harmful ways. This lack of trust becomes an integrity issue. Unfortunately, in some organizations, integrity is becoming scarce.

There is a secondary effect of our losing our integrity through the way in which we treat one another in the organization. It begins to impact the way we treat those outside of the organization; the ones we want to buy or secure our product or service. Maintaining integrity is important as we connect with people outside of the company. We should do what we say we are going to do and back up our product or service by making good on our promises.

No one came to know this any better than a man who began a company years ago named Leon Leonwood Bean. He started a mail order business in 1912 in Greenwood, Maine. His first product was a hunting boot that he sold with a money-back guarantee. This seemed to be a good plan until 90 percent of the products were returned because of a defective design. If he made good on his guarantee it could ruin him. Although this was a very real possibility, he kept his word, corrected the design, and continued to sell the boots. L.L. Bean is now one of the largest mail-order companies in the United Sates, in large part because it has continued to treat its customer's with integrity.[7] Where did this begin? With the leader of the organization!

Quality Four:
A Team Openly Exchanges Ideas

To secure the buy-in of people to an idea or to a mission, team members need to feel the freedom to share openly about how the organization can operate in the most effective way. Participants must buy-in to the purpose and plan if team success is to be realized. Those who share ideas and are appreciated for sharing them are willing to commit to organizational efforts. They know their opinion is valued. Let's take a look at what a team does to encourage an open exchange of ideas.

A team believes in the power of Collective IQ.

We learned in an earlier chapter about the power of Collective IQ which says that *we are smarter together than we are apart.* I am amazed at how often I have entered a room of leaders to share an idea I felt would help to improve our organization only to leave with a completely different plan. The reason is the freedom people have to share their thoughts about what is suggested. Partners have different perspectives that we often neglect to consider which have a bearing on the success of what we do.

A team follows the rule that leads to open communication.

When the proper environment has been set by establishing the rule that "there are no bad ideas," a dialogue can take place that strengthens a plan. It is strengthened because of changes that are made due to new perspectives. Organizational partners commit at a much higher level when they have ownership in the plan and are much less likely to criticize because of their involvement. Criticism typically comes from the spectators who are on the sidelines who have had no role to play in the game. Depending on where the criticism comes from, it can be a sign that we have not done a good enough job in including key players in the conversation.

A team leader understands the importance of having a starting point.

To encourage an open discussion, I have found it important to have a starting point. Leaders who are successful understand that what they suggest might not be the best idea, but it starts the conversation. The mark of a good leader is that he or she puts enough thought into an issue to offer suggestions that can lead to meaningful dialogue that produces worthwhile strategies and plans.

Team members understand that there are different teams.

We need to recognize our role. Partners have the responsibility to share ideas with the team on which they are a member. There are different types of teams. There are teams that work together to formulate the mission. There are other teams that work together to define the purposes and goals that need to be accomplished to fulfill the mission. There are yet other teams that work together to perform the tasks that accomplish the purposes which have been defined. In each case, they need to operate as a team. This means that in each environment there is a collective effort to help formulate decisions. If one person is making all of the decisions without input, this isn't a team.

Quality Five:
A Team Accepts Accountability

In our social media generation, the way we view friendships is changing and it is affecting how we interact within organizational life. William Deresiewicz examined the new forms of friendship that have emerged in the age of Facebook in an article for *The Chronicle of Higher Education.* He pointed out that while the social media has opened up the door of opportunity to be connected with everyone, it more often than not comes at the expense of deep, meaningful, shaping friendship. He wrote, "Concerning the moral content of classical friendship, its commitment to virtue and mutual improvement, that… has been lost. We have ceased to believe that a friend's highest purpose is to summon us to the good by offering moral advice and correction. We practice, instead, the nonjudgmental friendship of unconditional acceptance and support – "therapeutic" friendship, (to quote) Robert N. Bellah's scornful term. We seem to be terribly fragile now. A friend fulfills her duty, we suppose, by taking our side – validating our feelings, supporting our decisions, helping us to feel good about ourselves. We tell white lies, make

_____

excuses when a friend does something wrong, do what we can to keep the boat steady. We're busy people; want our friendships fun and friction-free..." He concluded, "Friendship is devolving, in other words, from a relationship to a feeling – from something people share to something each of us hugs privately to ourselves in the loneliness of our electronic caves."[8]

We need people who care about us. Some would say it's okay not to have friends in the work environment. My response is that we all should act friendly. This happens when we really care about one another. An important characteristic of this type of relationship is that we are willing to share information that we need to share, not only information that people want to hear.

Partners must be willing to be held accountable and to hold others accountable. This occurs when they are passionate about a common goal. It also happens when relationships are strong between co-workers. Those who care about one another are much more willing to share information with others when they are not living up to expectations. They are also more willing to accept criticism when they have failed themselves. Criticism and accountability often go together.

There are three types of criticism. They include constructive criticism, destructive criticism, and meaningless criticism. Let's deal with the latter first. Meaningless criticism is given when someone shares their opinion about something that doesn't really matter. I'll give you an example. One of my duties in our household is to make the bed. My wife is director over a large preschool and leaves the house at the crack of dawn to begin her work day. That leaves me with this "fun" responsibility. Believe me - I don't see it as fun at all. I confess, I have an attitude at times. Why? – you may ask! Here's the reason. There are nine pillows on our bed. As a guy, I only see a need for two – one for Jennifer and one for me. For some reason she thinks that the pillows add value to the room. That is her opinion. She has her opinion and I have mine. The truth is, neither one of us are right or wrong. They are merely our preferences. I could criticize her over what she likes, but believe me, it will do no good. The end result would be an argument that led to friction and a bed that would still have nine pillows on it. What does this teach us? There are some things that are better kept to ourselves.

The other two criticisms are important for us to consider. Constructive criticism helps to build something that we deem important and destructive criticism tears down what we think is not important. Our feeling about the importance of people makes the difference in whether or not we use constructive or destructive criticism. If they are important, we will make certain that what we say has a positive effect and leaves the person feeling valued. If the person is not important to us, we may want them to feel hurt or destroyed depending on how little we care for them.

Think about this on a personal level. When someone criticizes you, do you feel valued or devalued? If we feel valued, we know they care. If we feel devalued, we feel they don't. As leaders, the goal is always to provide criticism that is accepted. They will accept it if they know we have their best interests in mind and want to help them become the people they can be.

THE BIG ENDING

Are you a team or are you a group of individuals? The time has come to decide as a team to accomplish results that are worth your time and effort. Our investing in one another to assist in our improvement as people is worth it. Make the commitment right now to do what is necessary to develop your team into a crew rowing in the same direction to accomplish a mission of significance. If you are in leadership, commit to following a new definition of a leader...

Leader:
Someone who cares enough about others to help them care more for others.

Can you imagine what would happen if you truly had a concern for one another and led out of that concern? Your organization and our world would be a better place. It begins you. You can do it!

Questions

- Do team partners understand what the team is to accomplish? If so, why do you believe this?

- What is your team attempting to accomplish?

- What process does your organization use to define roles?

- Do you believe that people are in the right roles? If not, why do you feel this way?

- Is the leader of the team accepted by team partners? Why or why not?

- Are you accepted as a team leader? Why or why not?

- Do team partners understand the importance of the plan developed by the team? If so, what evidence do they give that reveals this?

- Do you have a plan to accomplish the team's purpose? If so, what is it?

- Are relationships between team partners causing a sense of unity or division? Why do you believe this to be true?

- How would you describe the effectiveness of communication in your organization?

- Who in the organization is having difficulty in relationships with others because of poor communication? Why is the communication poor?

- Is your team communicating its purpose, roles, expectations, and plan, as well as the personal needs of its partners? If so, how is the team accomplishing this?

Notes
(Other thoughts related to what you have learned in this chapter)

MY ACTION PLAN

• What will you do to begin having a better understanding
  of your organization?

- What will you do to build passion for your role within the organization?

- What will you do to build relationships with people within the organization?

- What will you do to better define the purpose of your organization?

- What will you do to develop or adjust your plan for accomplishing your purpose?

- What will you do to develop your leadership abilities in areas of weakness?

- What will you do to continue growing in your leadership in areas of strength?

- What will you do to develop your team?

END NOTES

INTRODUCTION

1. Peggy Noonan, *John Paul the Great* (Wheaton, IL: Viking, 2005), 110.
2. Jim Collins, *How the Mighty Fall* (New York: Harper Collins Publishers, 2009), 20-23.

1
THE PROPER MINDSET

1. Philip Yancey and Dr. Paul Brand, *In the Likeness of God* (Grand Rapids, MI: Zondervan 2004, 35.
2. Ibid., 36.
3. Ibid., 35-40.
4. Dennis Perkins, *Leading at the Edge* (AMACOM, 2000), xiii-xiv.
5. Mae Anderson, *"Toys 'R' Us opening 600 holiday stores in malls, hiring 10,000,"* USA Today (9-9-2010).

2
THE PASSION

1. Erwin Raphael McManus, *An Unstoppable Force* (Loveland, CO: Group Publishing, 2001), 14.
2. Mary Beth Marklein, *"Freshmen Have Making Money on Their Minds,"* USA Today (1-21-2010).
3. Jae Yang and Sam Ward, "Snapshots: Who appreciates you the most at work?" *USAToday.com* (7-10-2010).

4. Jim Herrington, Mike Bonem, and James Furr, *Leading Congregational Change* (San Francisco, CA: Jossey-Bass Publishing, 2000), 34.
5. John Kotter, *Leading Change* (Boston: Harvard Business School Press, 1996), 36.
6. Kotter, 21.

3
THE PURPOSE

1. Dan Buettner, *Costa Rica Secrets to a Long Life,* (AARP Magazine: May/June, 2008), 69.
2. McManus, *An Unstoppable Force*, 67.
3. Pat MacMillan, *The Performance Factor* (Nashville, TN: Broadman & Holman Publishers, 2001), 44.
4. Robert Krulwich, *Circling Themselves to Death,* Blog: 10:39 am, February 22,2011, www.npr.org/blogs/krulwich,2011/02/22.

4
THE PLAN

1. Peter Senge, *The Fifth Discipline* (Currency, 2006), 321.
2. David Guralnik, *Webster's New World Dictionary* (New York: Simon & Schuster Publishing, 1982), 570.

5
THE PERSON

1. John Maxwell, *Winning With People* (Nashville, TN: Nelson Books, 2004), 20.
2. Daniel Goleman, "Leadership That Gets Results," Harvard Business Review (March-April, 2001), 78.
3. Ibid., 82-83.
4. Charles Colson, *How Now Shall We Live* (Carol Stream, IL: Tyndale, 2004).
5. Michael Jordan, *Cyber Nation,* from the daily e-mail, *"Great Quote To Start Your Day on a Positive Note"* (5-7-2002).
6. Diandra Leslie-Pelecky, *The Physics of NASCAR* (Dutton, 2008); as seen in *Time* (3-3-08), 43.
7. Hal Bock, Associated Press, *"A Coach for All Seasons,"* The Spokane-Review Newspaper (December 4, 2000), c8.
8. Paul Coughlin, *"The Problem With Nice Guys,"* Focus On The Family Magazine (June, 2007), 8.
9. George Barna, *Grow Your Church From the Inside In* (Ventura, CA: Regal Books, 2002), 57.
10. Goleman, "Leadership That Gets Results," 80.
11. Ibid., 80.
12. Ibid.
13. Ibid.
14. Ibid.

6
THE PARTNERSHIP

1. David Waldstein, *"Keeper of Giants's Bats and of Team's History,"* The New York Times (10-3-10).
2. Theresa Kline, *Remaking Teams: The Revolutionary Research-Based Guide That Puts Theory Into Practice* (San Francisco: Jossey-Bass, 1999), 8-9.
3. MacMillan, *The Performance Factor*, 9.
4. Malcolm Gladwell, *Outliers* (Lebanon, IN: Little, Brown and Company, 2008), 115.
5. Fast Company Magazine, *First Impression Newsletter* (August 2, 2006).
6. Joe Posnanski, *"Made to Last,"* Sports Illustrated (August 23, 2010), 49-51.
7. Louis Upkins Jr., *Treat Me Like a Customer* (Grand Rapids, MI: Zondervan, 2009), 110.
8. William Deresiewicz, "Faux Friendship," *The Chronicles of Higher Education* (12-6-2009).